Network Programming and Automation Essentials

Get started in the realm of network automation
using Python and Go

Claus Töpke

BIRMINGHAM—MUMBAI

Network Programming and Automation Essentials

Group Product Manager: Mohd Riyan Khan
Publishing Product Manager: Mohd Riyan Khan
Senior Editor: Romy Dias
Technical Editor: Nithik Cheruvakodan
Copy Editor: Safis Editing
Project Coordinator: Ashwin Kharwa
Proofreader: Safis Editing
Indexer: Hemangini Bari
Production Designer: Vijay Kamble
Marketing Coordinator: Agnes D'souza

First published: March 2023
Production reference: 1100323

Published by Packt Publishing Ltd.
Livery Place
35 Livery Street
Birmingham
B3 2PB, UK.

ISBN 978-1-80323-366-6
www.packtpub.com

To all engineers that work hard to connect humans to network infrastructure by automating them.

Contributors

About the author

Claus Töpke is a product developer and founder of Telcomanager. He has worked with large network service providers, such as Telstra, NBN Australia, NZ Telecom, AWS Australia, AWS US, and Embratel. He has also worked in conjunction with large network technology corporations, such as Nokia, Amazon, Juniper, and Cisco. He has been able to experience different job titles, passing through fields such as network engineering, network performance, product development, and software engineering. His experience with network automation has led to the construction of several products and systems for different companies. He also worked on network performance for his master's thesis and wrote a book about service providers.

A special thanks to my wife and my whole family who have always supported me throughout the journey of this book. In addition, I must mention my little son Daniel, who tried very hard to get me away from the computer screen to play, an endeavor that was successful most of the time but was gunpowder for my motivation and inspiration. Also, a big thanks to Telcomanager for the support.

About the reviewers

Johan Lahti is a technology enthusiast with almost 15 years of experience working with computer networks. He has gained a wide range of experience including designing, building, and maintaining data center, enterprise, and service provider networks and automating those for various customers, ranging from ISPs to the public sector and large enterprises. He also teaches network and network automation for different educational institutes.

He is the founder and CEO of Acebit AB and works primarily as a senior consultant helping many different customers with design and architecture for network and infrastructure automation.

Radoslaw Majkut has worked for over 20 years in the computing technology field, including 13 years at Amazon and 3 years with his current employer, Google, in roles such as customer support, systems admin, SRE, and software engineer. He worked with Claus at Amazon on a team that developed a large-scale network simulator system.

Table of Contents

3

Accessing the Network   45

4

Working with Network Configurations and Definitions   73

Part 2: Network Programming for Automation

5

Dos and Don'ts for Network Programming   95

6

Using Go and Python for Network Programming   123

7

Error Handling and Logging   151

8

Scaling Your Code   171

Part 3: Testing, Hands-On, and Going Forward

9

Network Code Testing Framework   207

10

Hands-On and Going Forward   227

Preface

Howdy! As I transition from network engineering to software engineering, I understand how hard it is for developers to understand computer network concepts and how hard it is for network engineers to understand software concepts.

As applications become cloud-based and content is stored in the cloud, network infrastructure is the most important asset for today's businesses. A 10 GE interface down for 15 minutes is equivalent to 1 TB of missing data. Minimizing downtime, maximizing capacity and speed, and lowering latency and jitter are the major performance indicators for network providers today. And it's becoming impossible to improve these indicators without network automation, from the concept and design of the network to daily network operations.

This book will start with a brief description of network basics, then will start to explore network programming concepts and the automation ecosystem in a bit more depth. It will then describe protocols, tools, and techniques for network programming. All of these will be based on the two most popular computer languages for network automation, which are Go and Python.

Hands-on labs will be used to support the concepts covered in the book using a real-time network emulation. By the end of the book, you will have enough knowledge to start programming and creating network automation solutions.

Who this book is for

This book is for network architects, network engineers, and software professionals who wish to integrate programming into networks. Network engineers who follow traditional techniques can read this book to understand modern-day network automation and programming.

What this book covers

Chapter 1, *Network Basics for Development*, is focused on explaining the basics of computer networking and the jargon used. The idea is to build a good foundation to be built on throughout the book. If you are a network engineer or have experience in this field, you might want to skip it. If you are a software developer with little network experience, this chapter is for you. It will help you build a solid base of network jargon that will be useful when writing code for network automation.

Chapter 2, *Programmable Networks*, looks into several different technologies used today to create networks via software. Then we will examine the current standard technology, known as **Software Defined Networks (SDNs)**.

Chapter 3, Accessing the Network, explores the most common methods and protocols for accessing network devices for our network automation. Since devices have multiple methods, we will aim to give you enough information so that you can choose the one that is most appropriate for your network automation code.

Chapter 4, Working with Network Configurations and Definitions, explores how to work with a network configuration and how to define it for effective use with network automation. We want to build scalable and future-proof solutions. The chapter answers why we need network definitions to help automation.

Chapter 5, Dos and Don'ts for Network Programming, focuses on the most important best coding practices for Python and Go related to network programming. Writing code for networks is exciting because when it fails, it is challenging, and when it works, it is rewarding. If you are an experienced programmer, it will be plain sailing, but if you are a newbie, it will be stormy. The chapter dives into some coding practices for Go and Python that will help you get through these storms more easily.

Chapter 6, Using Go and Python for Network Programming, looks at how Python and Go are powerful and useful for network programming, but depending on what the requirements are and the environment, one might be better suited than the other. We'll also check the advantages and disadvantages of using Python and Go.

Chapter 7, Error Handling and Logging, explores how we report program execution events and how we handle errors. These two topics are not as easy as they seem, and they are, most of the time, implemented in systems poorly. The chapter investigates how and why we handle errors and then why and how we do event logging.

Chapter 8, Scaling Your Code, covers some techniques used today to scale your code up and down effectively, which will allow your solution to adapt easily to follow network growth, and, if necessary, easily scale down to save resources.

Chapter 9, Network Code Testing Framework, focuses on techniques to build a network testing framework that can be used for testing your network automation code. It also looks into advanced techniques that can be used to make your testing framework even more useful and reliable.

Chapter 10, Hands-On and Going Forward, explored building a network from scratch using the network automation skills we've learned and emulated routers. The finished emulated network will have enough components to experiment with several techniques described in the chapters of this book. And finally, there are a few remarks and some guidance for future studies and work.

To get the most out of this book

You will need to be able to work with privileged access with Linux, Windows, or macOS, and be able to install the Go and Python languages, as well as the corresponding third-party libraries. A basic understanding of computer networking concepts will also help.

Software/hardware covered in the book	Operating system requirements
Preferably Go language version 1.20 or later	Windows, macOS, or Linux
Preferably Python version 3.10.9 or later	Windows, macOS, or Linux

Download the example code files

You can download the example code files for this book from GitHub at `https://github.com/PacktPublishing/Network-Programming-and-Automation-Essentials`. If there's an update to the code, it will be updated in the GitHub repository.

We also have other code bundles from our rich catalog of books and videos available at `https://github.com/PacktPublishing/`. Check them out!

Download the color images

We also provide a PDF file that has color images of the screenshots and diagrams used in this book. You can download it here: `https://packt.link/AXHbe`.

Conventions used

There are a number of text conventions used throughout this book.

`Code in text`: Indicates code words in text, database table names, folder names, filenames, file extensions, pathnames, dummy URLs, user input, and Twitter handles. Here is an example: "Both modes use the same Linux device driver (accessible via `/dev/net/tun`), just with a different flag. The flag to use TAP mode is `IFF_TAP`, whereas the flag to use TUN is `IFF_TUN`."

A block of code is set as follows:

```
from paramiko import SSHClient

client = SSHClient()
client.connect('10.1.1.1', username='user', password='pw')
```

When we wish to draw your attention to a particular part of a code block, the relevant lines or items are set in bold:

```
import unittest
import mock
from paramiko import SSHClient

class TestSSHClient(unittest.TestCase):
    @mock.patch('paramiko.SSHClient.connect')
    def test_connect(self, mock_connect):
```

Any command-line input or output is written as follows:

```
claus@dev:~$ sudo ip tuntap add dev tap0 mode tap
claus@dev:~$ sudo ip link set tap0 up
claus@dev:~$ ip link show tap0
```

Bold: Indicates a new term, an important word, or words that you see onscreen. For instance, words in menus or dialog boxes appear in **bold**. Here is an example: "Having included **E5**, **S5**, and **L5**, the Clos network will have now 50 connections."

> **Tips or important notes**
> Appear like this.

Get in touch

Feedback from our readers is always welcome.

General feedback: If you have questions about any aspect of this book, email us at customercare@ packtpub.com and mention the book title in the subject of your message.

Errata: Although we have taken every care to ensure the accuracy of our content, mistakes do happen. If you have found a mistake in this book, we would be grateful if you would report this to us. Please visit www.packtpub.com/support/errata and fill in the form.

Piracy: If you come across any illegal copies of our works in any form on the internet, we would be grateful if you would provide us with the location address or website name. Please contact us at copyright@packtpub.com with a link to the material.

If you are interested in becoming an author: If there is a topic that you have expertise in and you are interested in either writing or contributing to a book, please visit authors.packtpub.com

Share your thoughts

Once you've read *Network Programming and Automation Essentials*, we'd love to hear your thoughts! Scan the QR code below to go straight to the Amazon review page for this book and share your feedback.

https://packt.link/r/1803233664

Your review is important to us and the tech community and will help us make sure we're delivering excellent quality content.

Download a free PDF copy of this book

Thanks for purchasing this book!

Do you like to read on the go but are unable to carry your print books everywhere? Is your eBook purchase not compatible with the device of your choice?

Don't worry, now with every Packt book you get a DRM-free PDF version of that book at no cost.

Read anywhere, any place, on any device. Search, copy, and paste code from your favorite technical books directly into your application.

The perks don't stop there, you can get exclusive access to discounts, newsletters, and great free content in your inbox daily

Follow these simple steps to get the benefits:

1. Scan the QR code or visit the link below

https://packt.link/free-ebook/9781803233666

2. Submit your proof of purchase
3. That's it! We'll send your free PDF and other benefits to your email directly

Part 1: Foundations for Network Automation

The first part is dedicated to refreshing you on some of the network fundamentals and jargon, as well as discussing some important aspects of network automation that should be used as the foundation of your work. Additional bases for network automation, such as the methods and protocols used to access the network and how we should use the network configuration and definitions, are also discussed in this part.

This part has the following chapters:

- *Chapter 1, Network Basics for Development*
- *Chapter 2, Programmable Networks*
- *Chapter 3, Accessing the Network*
- *Chapter 4, Working with Network Configurations and Definitions*

1

Network Basics for Development

This chapter is focused on explaining the basics and jargon used in **computer networking**. The idea is to build a good foundation to be used throughout the book.

If you are a *network engineer* or have experience in this field, you might want to skip it, or perhaps skim through it.

If you are a *software developer* with little network experience, this chapter is for you. It will help you build a solid base on network jargon that will be useful when writing code for **network automation**.

The following are the topics that we will cover in this chapter:

- Reviewing protocol layers, network device types, and network topologies
- Describing network architecture and its components
- Illustrating network management components, network bastions, and more

Reviewing protocol layers, network device types, and network topologies

We have lots to talk about here. But due to the size restraints of this book, I have organized a summary with the most important aspects of today's network jargon and explained them briefly. I hope you can find some new information to help your automation work.

Protocol layers

It's important to note that there are several different standards for protocol layers, and the most academic one is the ISO organization called **OSI model**, which defines seven layers. But we are going to consider only five defined in the TCP/IP protocol stack, which is used on the internet. Here is a short summary of each of the layers:

- **Physical layer**: In this layer are the technologies involved in the physical connection itself where the bits and bytes are transformed into the physical medium, such as the light in fiber optics, electricity in a cable, and radio waves in antennas. At this layer, physical checks can be implemented on the node input, such as power levels, collision, noise, and signal distortion, among other types of checks.

- **Data link layer**: Here, the information is called a **frame**, and it contains a delimited size, known as the **maximum transmission unit** (**MTU**). The reason is that a frame is a data representation in bytes that has to move from one node to another one and in a reliable manner without interruption. At this level, frame queues are present; the queues are used to place the frames on the physical layer in sequential order or in priority order. Some data link devices can prioritize certain types of frames, jumping to the front of the queue. At the data link layer, some checks are done, but within the frame itself, such as CRC or checksum. In addition, source and destination addresses can be added to the frame to differentiate destinations on a shared media. The information on the frame is normally used locally within the same organization. This layer is also known as the **Ethernet layer**.

- **Network layer**: This is also known as the **IP layer**, or the router layer. Here, the information is called a **packet**, and it contains the information that goes between nodes that are beyond the layer 2 domain (or the previous Ethernet layer). This level is where the routing protocols are used, the **network address translation** (**NAT**) does its job, some **access control lists** (**ACLs**) are present, and the control packets are, among other functions. The packet on this level has enough information to know where it came from and where it has to go. This layer is also responsible for fragmenting the packet into multiple packets if the frame MTU is smaller than the IP packet. The main information carried in the packet is the **IP address** and has source and destination addresses.

- **Transport layer**: The transport layer deals with data information that is called a **segment**. On today's internet, only two types of protocols are used here, the **User Data Protocol** (**UDP**) and the **Transmission Control Protocol** (**TCP**). The idea is one provides more confirmation and control than the other. TCP has traffic flow control, packet loss detection, and packet retransmission, among other functions. UDP, on the other hand, is just the IP packet plus a little more information. The idea behind having TCP is to enhance communication on the unreliable internet, so the application has a guaranteed transport method. TCP has more overhead, with an additional header field, and might be slower in some cases than UDP. The transport layer adds a **port** number to the segment, which is carried inside every packet in the IP layer. The port number is used for two reasons: to designate which application is using the transport layer, such as port 80 for HTTP communication, and to associate it with a communication **socket** in the host. The port number is required for the source and destination, which will be used to designate the correct socket to communicate with the host.

- **Application layer**: This is the top of the layers, normally referred to by my professor as the *cherry on the cake*. An application layer is used to associate a **socket** on the host where data will be sent and received. The application normally handles the content of the data, such as page requests on HTTP. The software that we are producing in this book uses this layer to automate the network.

LAN, WAN, internet, and intranet

LAN, or **local area network**, is used to refer to networks that are local. Nowadays, it means networks that use the data link layer as the main communication, such as Ethernet. The reason why the name is more related to the communication layer than the geography is that technology has evolved, allowing Ethernet switches to communicate over thousands of kilometers. So, a LAN normally designates a topology inside the same organization using Ethernet, but not necessarily geographically in the same location.

WAN, or **wide area network**, is used to refer to networks that are remotely connected, or technologies that allow nodes to be far apart, such as extinct technologies such as X.25, Frame Relay, and **Asynchronous Transfer Mode** (**ATM**). Now, the term WAN is normally used to designate interfaces or networks that are connected to different networks, or in other words, networks that are not in the same organization, data link layer, or Ethernet domain.

> **Information**
>
> For more information about ATM, please refer to the article *Technology and Applications* in SSRN Electronic Journal, June 1998, by Jeffrey Scott Ray.

The internet is what you know, this gigantic network interconnecting everybody worldwide.

The term intranet was used when corporations were using the internet protocols to communicate internally on their network. The reason is that other technologies were competing with the internet TCP/IP protocol at that time, such as SNA and IPX. So, when the term intranet was used, it was simply to state that the corporate network uses TCP/IP. Nowadays, intranet refers to a network that is within the same organization and not connected to external nodes. Therefore, the network is *safe* from external interference.

Point-to-point connections

A **point-to-point** (**P2P**) connection is used to interconnect two nodes. A link between two nodes is normally a P2P connection (as shown in *Figure 1.1*), unless using media such as satellite or broadcast antennas. This connection can either be *back to back* or not. The term *back to back* is normally used to indicate that the nodes are connected directly without any other physical layer between them, such as repeaters. Therefore, back-to-back connections have limited distances due to the noise and distortion introduced in the connection as the wiring gets longer. Depending on the speed and the technology used, the distances are limited to within the same room or building.

Figure 1.1 – A P2P connection

Star or hub-spoke topologies

Star or hub-spoke topologies are used in small and medium companies, where one office is the main distributor and the other locations are consumers. The topology looks like a star, and network elements are smaller and simpler at the remote locations, while being larger and complex at the main distributor (see the example in *Figure 1.2*).

Normally, these types of topologies can scale up to hundreds of nodes, but depending on the traffic, the requirements can scale to thousands. Let's look at two examples that illustrate the scale of these topologies.

For instance, in a bank, the automated teller machines are distributed in remote locations, where the main computer is located in the main branch. This can scale to thousands of remote machines as the traffic requirements are small in terms of byte transfer on a teller machine.

On the other hand, if you have a supermarket chain using a star topology, it won't scale to thousands of remote machines, as each supermarket requires a large amount of data transfer to handle all transactions and employees.

So, the use of star topologies is limited to the amount of traffic it can handle in the central node. In the star topology, we have two device functions, a device that will be either at the remote location or in the main office.

Network capacity planning is trivial when dealing with star topologies, as the main office node is updated as it grows.

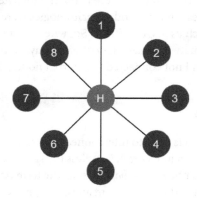

Figure 1.2 – A star topology

Hierarchical or tree topologies

Hierarchical topologies are used to optimize traffic, where larger nodes are used to aggregate traffic to smaller nodes in a hierarchical matter (see the example in *Figure 1.3*). These topologies can scale to thousands of nodes; however, because of the number of nodes in the path, the topologies can cause undesirable latency and extra node costs.

An internet service provider normally uses a hierarchical topology to concentrate customer traffic in certain remote locations before aggregating even more in other locations.

There is no limit on the number of nodes on this type of topology, and it's one of the foundations of the internet global infrastructure.

In the hierarchical topologies, we have multiple device functions, the **customer premises equipment** (**CPE**), aggregators, distributors, core, and peering, among others.

Depending on the size of this topology, it can introduce a longer path, which will add significant latency. For instance, in *Figure 1.3*, **A1** has to cross five hosts to reach **A7**.

Network capacity planning is focused on the aggregation points, and augmenting the network is not that difficult.

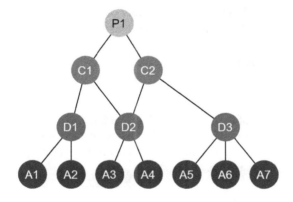

Figure 1.3 – A hierarchical or tree topology

Clos topologies

This type of topology is also known as a **Clos network** or **fabric**. This topology is used to increase the number of ports without compromising latency and throughput and is often used in data centers. This topology is composed of at least three stages. Note that there is no oversubscription or aggregation like in the hierarchical topologies. The Clos topology provides the same amount of available bandwidth on the input and output. The stage names are normally **spines** and **leafs**. The spines are always in the center and only have connections to the Clos nodes. Leaves are used to connect to external devices or networks.

Figure 1.4 shows an example of a 16-port Clos network. Note that normally, all connections between a spine node to a leaf node are *back to back*:

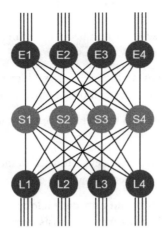

Figure 1.4 – A Clos topology

Why are these topologies used? To increase the number of ports available without compromising throughput. This kind of topology is also used inside a router to provide connectivity between interface cards. Some companies use small devices to increase the number of ports that are offered without raising the cost as smaller devices are normally cheaper.

> **Important note**
>
> One additional characteristic of the Clos network is that it has the same distance between any two external ports (in terms of nodes in the path), therefore the latency in normal conditions is the same. For instance, in *Figure 1.4*, the latency between an external port on node L1 to an external port on L4 or E1 is the same.

> **Important note**
>
> More information on Clos networks can be found in an interesting paper from Google called *Jupiter Rising: A Decade of Clos Topologies and Centralized Control in Google's Datacenter Network* – ACM SIGCOMM Computer Communication Review, Volume 45, Issue 4, October 2015.

Mixed topologies

A mixed topology is used in large corporations where latency and traffic are both important to care of. Normally, star topologies and P2P are used to shorten paths and reduce latency, whereas hierarchical topologies are used to optimize and aggregate traffic, and finally, Clos networks to increase the number of ports.

Modern cloud service providers are migrating to a more complex topology, where there are connections between elements where latency matters and aggregate device functions where traffic matters.

Network capacity planning is normally harder because connections are not totally hierarchical and aggregation points are not necessarily part of all traffic paths. An example of this kind of mixed topology is shown in *Figure 1.5*:

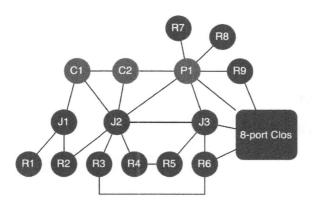

Figure 1.5 – A mixed topology

Interface speeds

A very important point that some engineers get confused about is the interface speed representation. 1 KB in memory representation is 2^10 or 1,024 bytes and 1 GB is 2^30, which is 1,073,741,824 bytes. For interface speeds, the same does not apply, and 1 Kbps is actually 1,000 bits/second, while 1 Gbps is 1,000,000,000 bits/second (more details can be found at `https://en.wikipedia.org/wiki/Data-rate_units`).

Device types and functions

Network devices used to have specific functions as CPU and memory were scarce and expensive. Nowadays, network devices can have multiple functions when required. In large networks, devices have fewer functions as they tend to get overloaded easier when traffic demands increase. Here are some of the functions that a device can have:

- **Hub**: This is a very old term to designate a device that only repeats the physical signal.
- **Switch**: A device that works only on the data link layer. It is normally used in LANs, and it works by switching frames. The most common protocol used on these devices to control paths is the **Spanning Tree Protocol (STP)**.

- **Router**: A device that works only on the network layer or IP Layer. It is used to interconnect multiple LANs or create long-haul remote connections. Internally, a router routes packets using a routing protocol to exchange route information with other routers. Some routers can also switch frames or work as a switch.

- **NAT**: NAT is devices that replace source and destination IP addresses to allow the use of private IP addresses or to isolate internal traffic from external traffic.

- **Firewall**: Normally, devices that control the traffic that passes through it by looking into the content of the frame or the packet. There are several different types of firewalls, and some might be super complex, which includes encrypting and decrypting traffic.

- **Load balancers**: When servers can't handle too many clients because of hardware limitations, load balancers can be used to deal with the client demands by sharing the client request between several servers. Those devices also look into the packet content to determine which server would get the traffic.

- **Network server**: A computer used to provide some sort of *service* to the network, for instance, an authentication server, an NTP server, or a Syslog collector.

Oversubscription

In network jargon, this term is used to describe nodes or links in the network that aggregate traffic from other parts of the network and statistically use it to their advantage. For instance, they have a 1 Gbps interface to connect to the internet and 1,000 customers with 10 Mbps interfaces to use the service, which is an oversubscription of 1 to 10. This practice is quite normal and is only possible to use because of the characteristics of the client's traffic that allow such aggregation without degradation. There are lots of mathematical models and papers on the internet describing this behavior and how to use it in your favor.

But some traffic can't be aggregated without being degraded. In a data center, the traffic that can't be oversubscribed is the traffic between servers, such as remote disk, data transfers, and database replicas. In this scenario, the best solution is to interconnect them without oversubscription using a solution such as non-blocking Clos topologies.

Browsing web pages, watching videos, and receiving messages from most of the traffic on the internet, which easily allows the aggregation technique without degradation.

> **Important note**
>
> More information on oversubscription can be found in the paper *Evaluating Impacts of Oversubscription on Future Internet Business Models* by A. Raju, V. Gonçalves, and P. Ballon – Published in Networking Workshops, 25 May 2012 – Computer Science.

In this section, we went over the basic components of computer networks, including protocols, topology types, interface speeds, and device types. By now, you should be able to identify these terms more easily and will be familiar with their meanings, because we are going to use these terms throughout this book. Moving on, we are going to review more terms related to network architecture.

Describing network architecture and its components

The term network architecture was introduced in the early 2000s, mimicking roles in the construction industry, where architects design and civil engineers build. Different companies use the term differently, but in this book, network architecture will be used to refer to the design of the network and its functions.

For a good network architecture, it is desirable to have a document describing in detail the first three layers of the network, from the physical layer to the routing layer. With this documentation, it is easy for the engineers to understand the physical connections, the Ethernet domains, and the routing protocols used.

Diagrams

A network diagram is mostly like a map, where the cities are the nodes and the roads are the links that connect them. For a network engineer, diagrams are crucial to describe how nodes are connected, and they also can group and demarcate important areas. A good diagram is easy to interpret and follow how data flows.

There are up to three types of diagrams; they can be integrated on the same page and graph, or they can be separated onto different pages. The main diagrams are one to show the physical connections, which can include the technology involved in the data link layer, and the switching and routing diagrams.

In *Figure 1.6*, we can see an example of a network diagram:

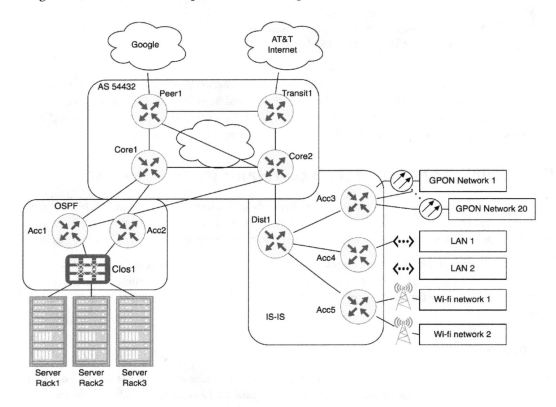

Figure 1.6 – Example of a network diagram

Figure 1.7 shows examples of network diagram symbols:

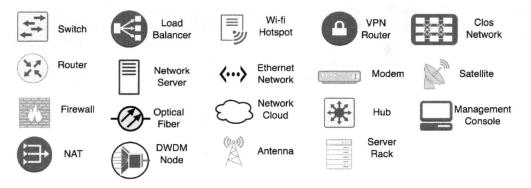

Figure 1.7 – Network diagram symbols

Network node names

A network node is a device that is essentially used to interconnect and serve as a transport of the data in the network. It can be either a hub, a switch, or a router. To help network engineers identify the node function, names are used to describe their main function. Here are some of them:

- **Transit router**: These are routers that have interfaces with other service providers. These links are normally used as a service to access other networks, therefore they have a cost because they are normally connected to other big carriers.

- **Peer router**: These routers have interfaces with other networks in a peer configuration, meaning none of the parts pay to use it. In these links, only the traffic between the peer companies is exchanged, and the traffic destined to outside networks are not allowed. Accessing external networks would be the case when using transit routers.

- **Core router**: These are nodes that are in the center of the network. They normally handle a large amount of traffic and have high-speed interfaces. Their throughput capacity is the highest in the network, but they have fewer interfaces as they concentrate the traffic of the network.

- **Distribution router**: These are nodes that normally connect to the core and aggregation routers. They normally interconnect different locations of the network. They don't have many interfaces and their throughput capacity is high, but not as high as the core router.

- **Aggregation router**: These routers normally aggregate the traffic from the access routers. They are normally located in the same area or location as the access routers, and they have fewer interfaces compared to the access routers.

- **Access router**: Some architects add a node that connects all last-mile networks or CPE nodes. These routers are located closer to the customer and have more interfaces than any other router.

- **Top of the rack (TOR)**: TOR refers to nodes that can be either a switch or a router, depending on the architecture. They are responsible for connecting the servers in the rack to the rest of the network.

- **Clos rack**: A Clos network, as described before, is a technique to add connectivity to multiple servers using small devices. A Clos rack is seen by the rest of the network as a single unique block, and in terms of architecture, it acts as a single node, normally used as a single router with a large number of interfaces.

- **CPE**: CPE is the node that is installed at the customer's location. It normally has one interface connecting to the last-mile network and one local interface that can be an Ethernet or a wireless Ethernet. These devices can also implement NAT, firewall and, in some cases, they have multiple local interfaces, which can act as a switch and a router. These nodes are cheap and small with very low throughput capacity compared to the other nodes.

The last-mile network

This term is used to describe the architecture used to connect the customer to the network. Normally, this term is only used for ISPs, but some corporations also use it to interconnect their branches.

The **last-mile network** has a range of coverage and normally doesn't cross the 1 km mark but depends on the type of technology used. Here are some of the most common last-mile networks:

- **Cable TV**: There are several technologies used here to provide data communication using the cable TV that the customer has installed. The most used one is DOCSIS, which in 2017 was upgraded to version 4. This solution uses a single cable that is shared to several premises.

- **Digital subscriber line (DSL)**: DSL uses the old telephone line to pass data communication. For that, there are lots of standards, and the most common ones are VDSL and ADSL. The DSL solutions don't share the same media as cable TV does, and there is one cable for each customer.

- **Fiber to the premises (FTTP)**: FTTP is when an optical cable arrives at a customer's premises. Like cable TV, the most common implementation is a single fiber that crosses several different customers in a sharable manner. The most common technology is a **passive optical network (PON)** or, more specifically, the **Gigabit Ethernet PON (GPON)** (or G.984).

> **Important note**
>
> Further details on GPON networks can be found in the paper *GPON in Telecommunication Network* – November 2010 – Paper from the International Congress on Ultra Modern Telecommunications and Control Systems (ICUMT) conference, 2010.

- **Wi-Fi**: Normally, this technology is used privately inside a company or a home, but some ISPs use the Wireless Ethernet standards (IEEE 802.11 family) to provide the last mile to customers using omnidirectional antennas. This particular use is different depending on each country and it depends on the government's legislation. They are normally advertised as **Ethernet hotspots** (https://en.wikipedia.org/wiki/Hotspot_(Wi-Fi)).

- **Satellite**: For data communication using satellites, there are two methods: one using geostationary satellites and the other using constellation satellites. The difference between them is the latency, as geostationary orbits very far from earth. The constellation method has low latency but has handover challenges as the satellites keep moving, normally having very low data throughput. The most famous technology using geostationary is VSAT. Internet using VSAT adds around 250 ms every time it has to travel from earth to the satellite, therefore it is a 500 ms round trip. But the dark ages of high latency might be over as SpaceX has announced they have finally solved the handover problem using the constellation method. This new service is called **Starlink** and has promised to have high capacity, low latency, and high availability using low orbit satellites.

> **Important note**
>
> A good discussion on the Starlink network can be found in the paper *Starlink Analysis* – July 15, 2021 – Research group ROADMAP-5G at the Carinthia University of Applied Sciences.

- **Power line communication (PLC) or HomePlug**: PLC, or **broadband over power lines (BoPL)**, uses the power cables to communicate data. This is achieved by modulating high frequencies on the wire. Most transformers won't be able to pass through the information as they act as a low-frequency cut filter, so it has to be contained within a house or between posts without a transformer. The most common technologies here are the HomePlug AV2 and IEEE 1901-2010 (`https://ieeexplore.ieee.org/document/5678772`).

- **Mobile**: Definitely the most popular network is the mobile last mile. Today, they use 5G technology, but other old networks are still in use, such as 4G (LTE), 3G, and GPRS.

> **Important note**
>
> More information on mobile technologies can be found at *Evolution of Mobile Communication Technology towards 5G Networks and Challenges* by A. Agarwal, K. Agarwal, S. Agarwal, and G. Misra – American Journal of Electrical and Electronic Engineering, 2019, Vol. 7, No. 2, pp. 34-37.

The physical architecture

The physical architecture is sometimes not necessarily the description of the cables or the fibers that will connect the devices but the infrastructure used by the network as a **physical** layer defined in the TCP/IP stack. This means we can reuse other foreign networks as a **physical layer** even though they have their own protocol stacks. Here are some of the possible physical technologies used in the architecture:

- **Dark fiber**: When connecting nodes, the term **dark fiber** means the nodes that are connected will be using a fiber that does not contain a repeater or underlying infrastructure. In the case of a connection between two nodes using dark fiber, if one node loses power, the other will not receive any light from the fiber. In this scenario, a fiber cut is perceived in both ends immediately, and interfaces go down instantaneously with a fiber cut. Only the packets in the output interface queue are discarded when a failure occurs.

- **Synchronous Transport Module (STM)**: STM was initially created to multiplex digital phone lines, but later started to be used for data communication. The most common one was STM-1, which was 155 Mbps. Routers used to have an interface that could encapsulates STM frames toward an STM network. The STM network would just switch the frames from one end to the other. A cut in the fiber using this technology might not be perceived quickly enough, causing a huge amount of packet loss. As we will describe later, **bidirectional forwarding detection (BFD)** needs to be used here to avoid drastic problems.

- **Dense wavelength-division multiplexing (DWDM)**: DWDM is an evolution of STM. The DWDM network is a switch network that also has a frame and time and wave division for each of the packets of data carried, similar to STM but enhanced. Similarly, BFD is necessary because a cut in the fiber here would not be perceived quickly enough, causing a huge amount of packet loss.

- **Back to back**: As explained before, the term *back to back* is normally used to designate the nodes that are connected directly without any other physical layer in between, such as repeaters.

- **Network tunnels**: Network tunnels are points of the network that are used to encapsulate the traffic and travel in a different network. Tunnels can be either Layer 2 or Layer 3 and are implemented to abstract the network that is being carried. In some network architectures, they are meant to reach a distant part of the network using a foreign infrastructure.

- **VPN tunnels**: These are like network tunnels. VPN tunnels normally add encryption.

The routing architecture

It's important to define how the traffic will flow in the network. For that, we need to have a proper design in terms of routing distribution. This is necessary so failure remediation, redundant paths, load balancing, routing policies, and traffic agreements can be implemented. The architecture would have to include an internal routing protocol and an external routing protocol if connected outside. Here is a summary:

- **Interior gateway protocol (IGP)**: IGP is a routing protocol that runs in a delimited area or location, normally internally within the same organization, as the name says. In the IGP domain, routers exchange path information by announcing and receiving topology updates. The most common IGPs use link state information to build the routing path topology. If an interface goes down, the update has to be propagated to the entire IGP domain. Isolated areas are used to avoid having to update a too-large topology and cause instability. Historically, the popular IGPs were RIP and EIGRP, but today, only **Open Shortest Path First (OSPF)** and **Intermediate System-to-Intermediate System (IS-IS)** are used.

- **Exterior gateway protocol (EGP)**: EGP is a routing protocol used to exchange routing information between organizations. It normally does not contain link state information, only the path distance. The most common EGP protocol is **Border Gateway Protocol (BGP)**.

- **IS-IS**: IS-IS is an IGP protocol designed by ISO, registered as ISO 10589. It is a link state protocol based on the shortest path algorithm called Dijkstra's algorithm. It's the second most used IGP.

- **OSPF**: OSPF is an IGP protocol designed by IETF, registered originally in 1989 by RFC1131 and updated a few times later. Version 3 is the last version described in RFC5340. OSPF also uses Dijkstra's algorithm to calculate paths and is the most popular and used IGP. OSPF uses areas to scale and improve stability during routing database updates.

- **BGP**: BGP is a unique protocol used to exchange routing information between organizations. It was first introduced in 1989 in RFC1105. It is also one of the protocols with more updates and extensions on the IETF and can be used for different purposes, such as **internal BGP (iBGP)**, **Multiprotocol BGP(MP-BGP)** defined in RFC4760, MPLS (MP-BGP), and recently, BGPsec, defined in 2017 in RFC8205. BGP is a path vector-based protocol, also known as a distance vector protocol, and it does not use link information like OSPF.

- **Autonomous system number** (**ASN**): Like the IP range, ASN is a unique number that is associated with an organization when starting using BGP to exchange routing tables. It is controlled by the five regional internet registries: **ARIN** in North America, **LACNIC** in Latin America, **APNIC** in Asia-Pacific, **RIPE** in Europe, and **AFRINIC** in Africa. When routing tables are exchanged using BGP, the ASN is carried on the path. For instance, Amazon.com uses ASN 16509 (`https://whois.arin.net/rest/asn/AS16509`).

Let's explore how a network works in terms of its state.

Types of failure

In computer networks, a major problem is the *instability* caused by failures in routing tables, links, or nodes. If a node goes numb, for example, the CPU freezes, the other nodes have to detect it quickly so they can divert the traffic through a different path. But how can a failure be detected to reroute quickly enough? Let's explore the types of failures first:

- **Link failure**: A link failure is when a connection between two nodes stops receiving or sending data because there is an interruption on the path. The failure can be caused by a physical problem, such as a fiber cut, environmental conditions, such as heavy rain, or because of middleware equipment failure. Nodes normally detect whether a link is down by the lack of signal on the input, but in some cases, such as when using repeaters or underlying networks (such as DWDM), the signal is present on the input but data can't be delivered. So, it requires a higher-level protocol to monitor and detect the communication breakdown instead of the interface input signal alone; otherwise, data will be discarded continuously until a node decides to reroute the traffic, which can take several seconds in some cases.

- **Node failure**: A node can fail in several different ways; the most common ones are power loss and OS freeze. A software glitch can cause a router to freeze for minutes or even hours, causing packet loss or not, depending on where the freeze occurs, in either the forwarding plane or the control plane. Detecting this failure quickly is a bit harder because all interface signals are still present, and the forwarding plane might be still working.

- **Flapping**: Interface flapping is when the interface keeps going down for short periods without being detected. Flapping causes data loss without detection and normally is hard to be discovered without specific equipment to measure the medium connected normally on both ends. The term **flapping** also is used when a route keeps appearing and disappearing on the routing table, called **route flapping**.

Failure detection techniques

Here are some techniques to detect failure:

- **Signal off**: Interfaces have a very simple way of detecting failure, by the absence of the main signal or light. In the case of fiber, if the intensity of the light received is too low, it would consider the interface down. Note that this detection is made on the input interface.

- **Protocol keep alive and hello packets**: Some routing protocols have keep alive (or *hello*) messages to check whether their neighbors are still alive. In OSPF, the default period for hello packets is 10 seconds for LAN interfaces, and 30 seconds for P2P connections. BGP has a default of 30 seconds. For today's network speed, 30 seconds is a lot of data lost. A 10 Gbps interface would discard a total of 37 GB if fully loaded. In today's protocol implementation, the period of sending these messages can't be shorter than a few seconds, which is still a long period of data lost.

- **Link BFD**: In 2010, IETF published RFC5880, which describes the BFD protocol, which was intended to allow routers to detect failure on their interfaces in the order of microseconds. The BFD message supports a minimum of 1 ms interval. BFD is normally implemented on the interface hardware, which allows it to respond without interrupting the main CPU.

- **The BFD routing protocol**: Link BFD is normally enabled in all interfaces of the network to detect failures quickly, but it would not help in the case of OS router freeze or control plane failure. To avoid packet loss in these cases, all major protocols have the BFD capability, including OSPF, IS-IS, and BGP. Although the BFD protocol message supports microsecond intervals, the implementation using routing protocols is normally in the order of milliseconds and limited to the number of points. The reason is that these messages need to be handled by the main CPU, and too many might cause performance degradation.

- **Route flapping detection**: The routing protocol can detect persistent route flapping and suppress it for a period. This is useful to avoid recalculating paths when a route is not actually stable. When suppression is in place, normally, the default route is taken.

Control plane and forwarding plane

It is very important to understand the difference between a forwarding plane and a control plane, especially if you are working on network automation. Let's explore them in this section.

The forwarding plane, or data plane, is an abstract concept where some processes, equipment, and hardware are used to forward traffic through the network. In other words, the forwarding plane defines all entities in the network responsible for receiving data, transporting it, and delivering it.

The control plane is an abstract concept designated to all entities in the network responsible for constructing the data path, removing it, or updating it.

A forwarding plane works when data is carried from one input point, *A*, to another output point, *B*, but does not need to have a control plane working. The control plane would only work if a path does not exist from *A* to *B*. The control plane also works in case of a failure because the original path might be interrupted and needs to be constructed again.

So, why is this important in network automation? Because the control plane has to update forwarding paths if there is a problem with the forwarding plane, which can cause packet drop, jitter, and delays. A stable network does not require any path updates and consequently minimum work for the control plane. Network automation needs to avoid any particular automation that might cause the control plane to update the network.

Graceful restart

Usually, when a router restarts, all the routing peers detect that the session went down and then came up. This *down/up* transition results in the control plane working to recompute all the route paths, generating thousands of updates in the entire network and, consequently, causing a churn to the forwarding plane. This recomputation can also cause routing flaps, which may create transient forwarding black holes and transient forwarding loops. These transient problems also consume a lot of resources on the control plane of the routers affected.

Therefore, a **graceful restart** was created to avoid such drastic changes if a restart is required.

The idea is we could restart all control plane processes in one router without affecting the forwarding plane and the control plane of the other neighbor routers. In practice, a graceful restart is a method to restart the routing processes without affecting the forwarding plane.

In 2003, IETF published RFC3623 to define the implementation of the graceful restart for OSPF. Today, the main control plane protocols have some sort of graceful restart, including BGP, IS-IS, MPLS, RSVP, and LDP.

When building network automation, this kind of method is preferred to update the software.

In this section, we've reviewed network architecture and its components. We got more details on routing and physical architecture components. We also learned how important control and data plane separation is, along with the failure types. It is important to know these network terminologies to help with network automation. Next, we're going to review network management and its components.

Illustrating network management components, network bastions, and more

Before we finish this chapter, let's touch on some of the terms used in network management and planning.

ACL

An **ACL** is used almost everywhere in the network to control access by filtering the IP packet based on the IP or port number.

The ACL can be implemented on either the forwarding plane or the control plane. When implemented on the forwarding plane, they are used to limit the IP reachability to certain parts of the network and also avoid IP spoofing. When implemented on the control plane, they are protecting the routing protocol and management ports from malicious connections.

ACLs are also used in the management interfaces to avoid undesirable traffic when using in-band management.

Management system and managed elements

The **management system** is the platform including the software and hardware responsible for managing the network. It can be centralized or distributed. The managed elements are the targets of the management system, including routers, switches, modems, repeaters, and intelligent racks, among others.

Note that the **managed element** does not need to be part of the network. It can be a support system, such as a rack with fans or an air conditioning unit. When writing code for network management, it is important to classify the elements appropriately so they can be managed accordingly.

In-band and out-of-band management

In-band and **out-of-band** (**OOB**) management refer to how the management system traffic reaches the management elements.

For OOB management, there is an isolated network infrastructure that carries only management traffic, which is not connected to the main network in any way and only to the management ports of every managed element. In other words, the forwarding plane does not carry any management traffic. In addition, the OOB network should be able to exist and deliver management traffic independently of the main network routing status, because, in the case of a catastrophic scenario, the OOB network should be enough to reach the managed element even if its network interfaces are down. It's important to note that this network normally does not carry much traffic and the nodes and interfaces on it have low throughput. Some OOB networks are implemented using mobile networks.

In in-band management, there is no physical network isolation between the forwarding plane and management traffic, so the interface that carries customer data also carries management traffic. In this scenario, ACLs are used extensively to avoid unwanted traffic toward the ports of the managed elements. In addition, some network architecture adds priority queues to interfaces to allow the management traffic to be delivered first and avoid discards on heavily loaded links.

Some management systems use both to talk to the devices, some via in-band and some via OOB networks. Normally, heavy traffic, such as OS upgrades and event logging, goes via in-band and the element console access goes via the OOB network.

Network telemetry

Telemetry is not a new term and refers to any type of equipment that can monitor field variables remotely, such as temperature or humidity. This term was then imported to computer networks to refer to a group of procedures used to collect network information remotely.

Network telemetry refers to an area in a computer network responsible for encompassing various procedures and systems to define, collect, and analyze network data. In some cases, it can mean a new method of obtaining network data by using streaming methods from the network devices.

Management information base

A **management information base** (**MIB**) is a formal description of a set of objects that can be managed using the **Simple Network Management Protocol** (**SNMP**).

An MIB can be public or private. When public, its definitions are published by an RFC, such as the interface group MIB defined on RFC2863. When private, it has to be provided by the vendor who owns the MIB.

The MIB is normally organized in a tree by numbers (*Figure 1.8*). When describing an object in an MIB tree, it is normally referred to as an **object identifier** (**OID**). For instance, the number of packets seen in an interface output is represented by an OID called **IfOutUcastPkts**, which has the sequence *.1.3.6.1.2.1.2.17* (http://www.net-snmp.org/docs/mibs/interfaces.html).

The OID normally contains a value that is a variable that can have different types, such as COUNTER, GAUGE, INTEGER, and OCTETSTR, among others.

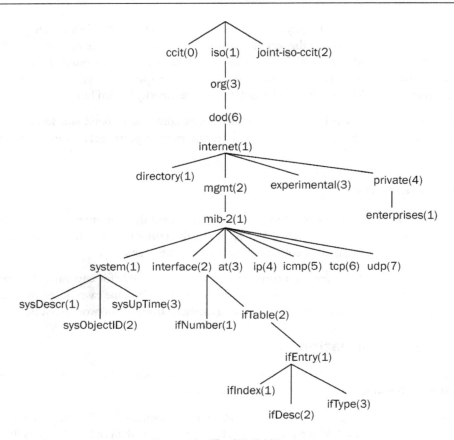

Figure 1.8 – The SNMP MIB tree

Network bastions

The term **bastion** comes from fortifications that were used to protect cannons during medieval wars. At that time, a bastion was an angularly shaped part of an outer wall, usually placed around the corners of a fort to allow defensive fire in many directions.

Like cannons in medieval times, network elements need layers for protection. **Network bastions** or **bastion hosts** are physical devices, normally computers, with defensive mechanisms built in that are connected to two or more networks.

Bastion hosts are popularly designed using Linux installed on a computer with multiple Ethernet ports. Each Ethernet port connects to a different part of the network where isolation or protection is desired.

To protect the networks, bastion hosts do not forward traffic, and normally, the IP forwarding capability is disabled like the following example applied in Linux:

```
# sysctl net.ipv4.ip_forward
net.ipv4.ip_forward = 0
```

A Linux box without IP forwarding means traffic can't be routed between interfaces, so traffic has to be originated in the bastion host to reach an external interface, and no other traffic would go out of the Ethernet port that is not originated locally in the box. Therefore, bastion hosts need to have authentication, such as a username and password, to allow a *user* or *system* to log in and run a shell locally. From the host, the user might be able to generate IP packets toward the other Ethernet ports.

Network automation will require an additional mechanism to allow accessing the network nodes for configuring them. We are going to cover these mechanisms later when writing some code to access the nodes.

An example of a bastion host is shown in *Figure 1.9*:

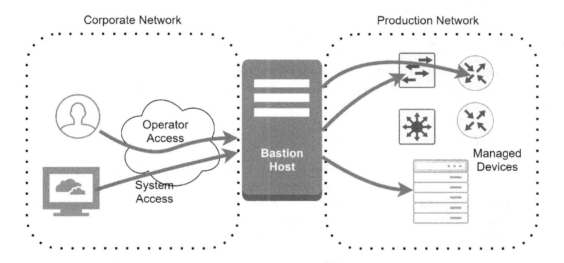

Figure 1.9 – Bastion host connecting production and corporate networks

FCAPS

FCAPS is a management network model and framework defined by ISO for network management. Its acronym stands for **fault, configuration, accounting, performance, and security**. Each of these is the management category into which the ISO model defines network management tasks. Let's look at a very summarized description of each of these tasks:

- **Fault**: The goal of fault management is to recognize, isolate, correct, and log faults that occur in the network.
- **Configuration**: This refers to storing configurations from devices, tracking changes, and provisioning new configurations.
- **Accounting**: This is concerned with tracking network usage for data transported, per business, client, or user. The goal is to be able to appropriately be billed or charged for accounting purposes.
- **Performance**: This is focused on ensuring that the network works at an acceptable level by monitoring network latency, packet loss, link utilization, packet discards, retransmissions, and error rates.
- **Security**: This refers to controlling access to assets and protocols in the network, which can include AAA systems, ACLs, and firewalls.

Why is FCAPS important for network automation? Because it's better to write network automation code with tasks separated like in FCAPS, so we can place automation in different parts of the management system accordingly.

Network planning

Growing the network is hard because if you buy too many resources, you could lose money. However, if you buy too little, you could lose customers. Network planning is used to ensure performance and costs are going in the right direction. In other words, it consists of several activities whose final target is to define an optimal cost-effective network design for the future. Network planning engineers work with prediction and statistical models to draw the most probable future network growth.

Because of the nature of the predictions, the network planning team needs a lot of data from the network, and most of them are probably not easy to collect. You might be required to collect instantaneous data, which is not available on the current management systems, so you can use your network automation skills for that.

Network security

One important point to mention is the work that companies are doing to create a safe place for traffic to flow. Most companies today have invested in separate and specialized teams to deal with security, and those teams have engineers that only understand nuances of the network to delineate security rules to be taken when designing and operating computer networks.

Your automation work will probably involve dealing with such security rules applied in the network, and will probably vary by company depending on the technology and level of security required. A good overview of network security can be found at `https://en.wikipedia.org/wiki/Network_security`.

Summary

In this chapter, we reviewed the key points of networking. The intention was to highlight the main concepts and define some of the network jargon. At this point, you have enough network background to discuss automation work with any network engineer. I hope that any work done on network automation coding from now on will make more sense and you're more familiar with the network terms.

In the next chapter, we will learn about how networks are evolving to be programmable.

2
Programmable Networks

Initially, computer networks were something physical and static, with wires and hardware, but with advanced computation, virtualization, and connectivity, networks have become more flexible and configurable by software. In this chapter, we're going to talk about how software has changed the picture for computer networks. We are going first to examine several different technologies used today to create networks via software, then we are going to examine the current standard technology, known as **software-defined networks (SDNs)**.

As we saw in the first chapter, computer networks can be quite complex and difficult to maintain. There are several different sets of equipment that range from routers, switches, and NATs to load balancers and more. In addition, within each piece of equipment, there are several different types of operation, such as *core* or *access* routers. Network equipment is typically configured individually with interfaces that are very different between each vendor. Although there are management systems that can help centralize the configuration in one single place, network equipment is configured at an individual level. This kind of operation means complexity in operation and slow innovation for new features.

In this chapter, we are going to cover the following topics:

- Exploring the history of programmable networks and looking at those used in the present day
- Virtual network technologies
- SDNs and OpenFlow
- Understanding cloud computing
- Using OpenStack for networking

Exploring the history of programmable networks and looking at those used in the present day

Several years have passed since **programmable networks** were initially conceived by engineers, so let's touch on a few historical milestones before we get into the current technologies.

Active networking

The **Defense Advanced Research Projects Agency (DARPA)** began funding research in the mid-1990s to create a network that could easily be changed and customized by programming, called the **active networking** project. The main goal of the project was to create network technologies that, in contrast to then-current networks, were easy to innovate and evolve, allowing fast application and protocol development.

But it was not easy to create such a flexible network in the 1990s because programming languages, signaling and network protocols, and operating systems were not mature enough to accommodate such innovative ideas. For instance, operation systems were monolithic and adding features required recompilation and reboot. In addition, service APIs were non-existent and distributed programming languages were still in early development.

The active networking research programs explored radical alternatives to the services provided by the traditional internet stack via IP. Examples of this work can be found in projects such as **Global Environment for Network Innovations (GENIs)**, which can be viewed at https://www.geni.net/, **Nacional Science Foundation (NSF)**, **Future Internet Design (FIND)** which can be viewed at http://www.nets-find.net/, and **Future Internet Research and Experimentation Initiative (EU FIRE)**.

At the time, the active networking research community pursued two programming models:

- The **capsule model**, where the code to execute at the nodes was carried in-band in data packets

- The **programmable router/switch model**, where the code to execute at the nodes was established by out-of-band mechanisms

> **Important note**
>
> Further reading can be found in *Active Networking – One View of the Past, Present, and Future.* Jonathan M. Smith and Scott M. Nettles – *IEEE TRANSACTIONS ON SYSTEMS - PART C: Application and Reviews, Vol 34 No 1, February 2004.*

Let's dive into one of the first attempts at creating a node that was programmable, known as **NodeOS**.

NodeOS

One of the first goals of the active networking project was to create **NodeOS**. It was an operating system whose primary purpose was to support packet forwarding in an active network. NodeOS ran at the lowest level in an active node and multiplexed the node resources, such as memory and CPU, among the packet flows that traverse the node. NodeOS provided several important services to active network execution environments, including resource scheduling and accounting and fast packet input-output. The two important design milestones on NodeOS were the creation of **application programming interfaces (APIs)** and resource management.

> **Important note**
>
> Additional reading on NodeOS can be found at *An OS Interface for Active Routers April 2001 – IEEE Journal on Selected Areas in Communications 19(3):473 – 487.*

Following this, we are now going to explore a few projects that were the early attempts from the community toward SDN.

Data and control plane separation

One of the major steps toward programmable networks and **SDNs** was the separation of the **control** and **data** planes. In *Chapter 1*, we discussed the difference between the control and data planes, and here we are going to discuss a bit of the history behind them. It's worth remembering that the data plane is also known as the **forwarding plane**.

By the 1990s, such separation was already present on public telephone networks but was not yet implemented in computer networks or on the internet. As network complexity increased and internet services started to become the main revenue for several backbone providers, reliability, predictability, and performance were key points for network operators to seek better approaches for managing networks.

In the early 2000s, a community of researchers who either worked for or regularly interacted with network operators started to explore pragmatic approaches using either standard protocols or other imminent technologies that were just about to become deployable. At that time, routers and switches had tight integration between the control and forwarding planes. This coupling made various network management tasks difficult, such as debugging configuration problems and controlling routing behavior. To address these challenges, various efforts to separate the forwarding and control planes began to emerge. Let's explore a few of the earlier efforts in the following sections.

IETF ForCES

Forwarding and Control Element Separation (**IETF ForCES**) working groups intended to create a framework, a list of requirements, a solution protocol, a logical function block library, and other associated documents in support of data and control element separation (`https://datatracker.ietf.org/wg/forces/about/`).

The NetLink interface

NetLink was perhaps the clearest separation of the control plane and data plane on the Linux kernel. In 2003, IETF published the *RFC3549* describing the separation of the **control plane components** (**CPCs**) and the **forwarding engine components** (**FECs**). *Figure 2.1* (from the original RFC) illustrates how Linux was using Netlink as the main separator between the control and data planes (`https://datatracker.ietf.org/doc/html/rfc3549`).

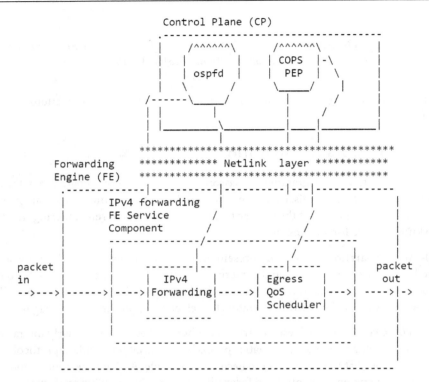

```
                    Control Plane (CP)
                  .-------------------------------------.
                  |    /^^^^^^\        /^^^^^^\          |
                  |    |       |       | COPS  |-\       |
                  |    | ospfd |       |  PEP  |  \      |
                  |    \       /       _____/   |     |
                  /------\____/            |        /    |
                  |  |           |         |       /     |
                  |  |_____|_____|_____|
                  |  |           |         |      |
                  ****************************************************
Forwarding        *************** Netlink  layer ***********
Engine (FE)       ****************************************************
          .-------------|----------|----------|---|--------------
          |             | IPv4 forwarding   |       |              |
          |             | FE Service        /       /              |
          |             | Component        /       /               |
          |             --------------/-------------/--------       |
          |             |            |           /          |       |
packet    |             |    --------|--      ----|-----    | packet|
in        |             |    | IPv4     |     | Egress  |   | out   |
-->--->|------->|---->|Forwarding|----->| QoS    |--->|   ---->|->
          |             |    |          |     | Scheduler|   |       |
          |             |    ----------       ----------    |       |
          |             |                                    |       |
          |             --------------------------------------       |
          |                                                          |
          -----------------------------------------------------------
```

Figure 2.1 – Control and data plane separation, as shown in RFC3549

Netlink was first created on series 2.0 of the Linux kernel.

Routing Control Platform

Routing Control Platform (**RCP**) is a pragmatic design to separate the control and data planes. The idea was to create a centralized control where all routing information was collected and then run an algorithm to select the best routing path for each of the routers of the network.

RCP was implemented by collecting routing tables from external and internal **Border Gateway Protocol** (**BGP**) from the routers in the current network, using this information in a centralized manner to choose the best path to each of the routers. With this approach, it was possible to leverage the existing network devices and have control plane and data plane separation.

> **Important note**
> More about RCP using BGP can be found in the paper *Design and implementation of a routing control platform* – Authors: Matthew Caesar, Donald Caldwell, Nick Feamster, Jennifer Rexford, Aman Shaikh, Jacobus van der Merwe – *NSDI'05: Proceedings of the 2nd conference on Symposium on Networked Systems Design & Implementation – Volume 2 May 2005 Pages 15–28.*

SoftRouter

The **SoftRouter** idea was presented at a conference in 2004 and patented in 2005. Again, the architecture had separation between control plane functions and data plane functions.

All control plane functions were implemented on general-purpose servers called **control elements** (**CEs**), which may be multiple hops away from the **forwarding elements** (**FEs**). There were two main types of network entities in the SoftRouter architecture, which were the FEs and CEs. Together, they constituted a **network element** (**NE**) router. The key difference from a traditional router was the absence of any control logic (such as OSPF or BGP) running locally. Instead, the control logic was hosted remotely.

> **Important note**
>
> More details on SoftRouter can be found in the original 2004 paper *The SoftRouter Architecture – T. V. Lakshman, T. Nandagopal, R. Ramjee, K. Sabnani, T. Woo – Bell Laboratories, Lucent Technologies, ACM HOTNETS - January 2004.*

The path computation element architecture

In 2006, the IETF Network Working Group published an RFC describing an architecture of a centralized controlled entity to make route path decisions, which they called the **path computation element** (**PCE**) architecture.

Initially, PCE architecture was invented to solve a problem in **multiprotocol label switching** (**MPLS**) where the **label switch path** (**LSP**) calculations were becoming very slow and heavy for each router to calculate. It was designed to do the calculations on a server inside or outside the network instead.

> **Important note**
>
> More details on PCE can be found in *RFC4655:* `https://datatracker.ietf.org/doc/html/rfc4655`

Let's now look at the most important project of all, which was the work that went toward OpenFlow and SDNs.

Ethane

Ethane was one of the most important projects and culminated in the creation of OpenFlow and SDNs. Initially, it was just a project from a PhD student that defined a network as a group of data flows and network policies to control the traffic, which is another way to see the separation between the data plane and the control plane.

The Ethane project had the idea of centralizing all network policies in one place. A new device joining the Ethane network should have all its communication turned off by default. The new device should get explicit permissions from the centralized server before connecting and its data flow should only be allowed on the permitted paths.

> **Important note**
>
> More on the Ethane project can be found in the 2007 original paper *Ethane: taking control of the enterprise* – Authors: M. Casado, M. J. Freedman, J. Pettit, J. Luo, N. McKeown, Scott Shenker – *SIGCOMM '07: Proceedings of the 2007 conference, pages 1–12.*

In this section, we explored a bit of the history behind programmable networks. We also explored a few of the main projects that led to the separation of the control and data planes, which was an important milestone toward SDNs. You should now be able to identify the significance of the separation and why it happened.

Virtual network technologies

Network virtualization is when software acts like network hardware and it is accomplished by using logically simulated hardware platforms.

Virtualization of networks is not a new concept, and we can find one of the first implementations in the mid-1970s with virtual circuits on X.25 networks. Later, other technologies also started using virtual concepts, such as Frame Relay and ATM, but they are now obsolete.

Loopback interfaces were based on electronics where loopbacks are used to create electric loops for the signal to return to its source for testing purposes. In 1981, the IETF referred to the reserved address range `127.rrr.rrr.rrr` with **RFC790**, which also outlined 32-bit IP address space classes. In 1986, with **RFC990**, the address range `127.rrr.rrr.rrr` was officially called **loopback**. Today, this IP range is used in computer platforms to designate the localhost IP address of the computer when using the TCP/IP stack (for example, `127.0.0.1`)

Another early implementation of network virtualization was the **virtual LAN** or **VLAN**. By 1981, David Sincoskie was testing segmenting voice-over-Ethernet networks to facilitate fault tolerance, something similar to what VLAN does. However, it was only after 17 years that, in 1998, VLAN was finally published as a standard by IEEE by the name *802.1Q*. By the 2000s, switched networks dominated the landscape with switches, repeaters, and bridges, making VLANs commonplace. A LAN without a VLAN is virtually impossible today.

There are several other network virtualization technologies that are used today. Let's explore the important ones in the following sections.

Virtual private networks

This is the concept of creating an isolated secured network overlay that is implemented on network carriers, service providers, and over the internet.

In other words, **virtual private network (VPN)** is a generic term that describes the use of public or private networks to create groups of users that are isolated from other network users, allowing them to communicate between themselves as if they were on a private network.

VPNs use end-to-end traffic encryption to enhance data separation, especially when using public networks, but this is not necessarily the case for all implementations. For instance, when using VPNs in MPLS networks, the traffic is not encrypted as it runs over private domains, and data separation exists only by packet encapsulation.

VPN is a generic name, but more specific names can be found, such as L3VPN, L2VPN, VPLS, Pseudo Wires, and VLLS, among others.

> **Important note**
>
> More on VPN and all related families can be found at `https://datatracker.ietf.org/doc/html/rfc2764` and `https://datatracker.ietf.org/doc/html/rfc4026`.

The VLAN was perhaps one of the most important virtualizations created in L2 networks. Let's now look at an interesting virtualization created for router gateways.

The Virtual Router Redundancy Protocol

This protocol was initially created by Cisco in 1998 with the name **Hot Standby Router Protocol (HSRP)**, defined in *RFC2281*. As the use of HSRP was very popular at the time, the IETF Network Working Group created the **Virtual Router Redundancy Protocol (VRRP)** (*RFC3768*).

The concept is simple, giving computers only one default gateway on their routing table by acquiring automatically using DHCP or by configuring manually. To use two routers redundantly, you might need to update all computers or use VRRP.

VRRP uses a virtual Ethernet address to associate with an IP address; this IP address is the default gateway to all computers on the network. *Figure 2.2* illustrates **Router A** and **Router B**, which can assume the IP address 10.0.0.1 using a virtual MAC address that is associated with both routers.

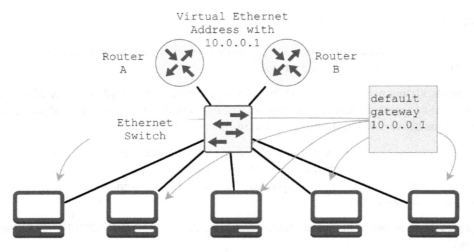

Figure 2.2 – VRRP using a virtual Ethernet address as the default gateway

> **Important note**
> More details on VRRP and HSRP can be found at https://datatracker.ietf.org/ doc/html/rfc2281 and https://datatracker.ietf.org/doc/html/rfc3768.

VLANs were created a long time ago, but its concept was used to extend to a more flexible usage, as we are going to see in the next section.

The Virtual Extensible Local Area Network

Perhaps the most important of all in virtualization today is the **Virtual Extensible Local Area Network** (**VXLAN**). This standard was published in 2014 and is heavily used for network virtualization to provide connectivity. With VXLANs, it's possible to create a network with interfaces connected back-to-back to routers like they are physical entities, but in reality, they are virtual.

A VXLAN encapsulates data link layer Ethernet frames (layer 2) within the transport layer using UDP datagrams (layer 4). VXLAN endpoints, which terminate VXLAN tunnels and may be either virtual or physical switch ports, are known as **Virtual Tunnel Endpoints** (**VTEPs**).

> **Important note**
> More about VXLANs can be found at https://datatracker.ietf.org/doc/html/ rfc7348.

Let's now explore an open source project that puts in place several virtual network technologies, including VLANs, VRRP, and VXLANs.

Open vSwitch

This open source project is perhaps the most important in network virtualization today. **Open vSwitch (OVS)** runs on any Linux-based virtualization platform (kernel 3.10 and newer) and is used to create connectivity in virtual and physical environments. The majority of the code is written in C, and it supports several protocols including VXLAN, IPSEC, and GRE, among others. OVS is an OpenStack component of SDNs and perhaps the most popular implementation of OpenFlow. A basic architecture of how OVS works can be found in *Figure 2.3*.

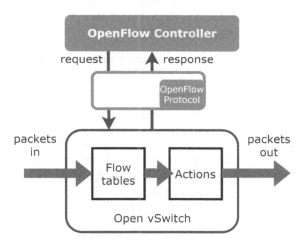

Figure 2.3 – Simplified OVS architecture

More details on OVS can be found at `https://github.com/openvswitch/ovs.git`.

Linux Containers

Linux Containers (LXC) provides operating-system-level virtualization using CPU, network, memory, and I/O space isolation. Its first implementation was on Linux kernel 2.6.24 in January 2008, but the concept is old and can be found in a FreeBSD implementation called **jails** implemented in 1999 and published on FreeBSD 4.0 in March 2000 (details at: docs.freebsd.org/en/books/handbook/jails/).

Today, more and more implementations of LXC can be found, but the concept of CPU, network, memory, and I/O space isolation is the same. The most popular LXC implementation today is **Docker**.

With LXC and Open vSwitch, it's possible to create an entire virtual network topology with hundreds of routers. A powerful example is **Mininet** (`http://mininet.org/` and `https://github.com/mininet/mininet`).

> **Important note**
> More on LXC and FreeBSD jail can be found at `https://en.wikipedia.org/wiki/` `LXC` and `https://en.wikipedia.org/wiki/FreeBSD_jail`.

Containers for Linux can create most virtualizations, however they are limited by using the same operational system because containers share the same kernel. Virtual machines, as we'll see next, can be used to virtualize a wide range of other operating systems.

Virtual machines

LXC is powerful in isolating parts of the operating system; however, they aren't able to run applications that require a different CPU or hardware. So, **virtual machines** (**VMs**) are there to add this extra virtualization by simulating physical hardware and CPU.

A VM can further isolate the operating system by creating a whole new layer of CPU, I/O, memory, and network. For instance, in network virtualization, it's possible to run different operating systems with different CPUs, such as Juniper JunOS using Intel CPUs, and Cisco IOS using MIPS CPUs.

The most popular open source implementation of VMs is **Xen** (`https://xenproject.org/`).

We do have much more to talk about regarding network virtualization, but that would be a topic for another book. At least for the time being, what we have examined in this section is sufficient to identify the main technologies used by programmable networks. At this point, you should be able to identify these technologies easily if you encounter them.

SDNs and OpenFlow

We have investigated a few historical milestones of programmable networks and network virtualization that form the base of what we know today as SDNs. Next, let's talk about the details behind SDNs.

In order for SDNs to be successful, they need to be flexible and programmable, making it simple to deploy and control traffic and manage their components. None of this could be done without separation between the control plane and the forwarding plane (the data plane).

SDN implementation is done by having an application that uses the decoupling of these two planes to construct the data flows of the network. This application can run in a network server or in a VM, which sends control packets to the network devices using an OpenFlow protocol when possible.

History of OpenFlow

OpenFlow is a standard protocol used in SDNs. Its origins can be traced back to 2006 with the project mentioned earlier in this chapter called Ethane. Eventually, the Ethane project led to what became known as OpenFlow, thanks to a joint research effort by teams at Stanford and Berkeley universities.

The initial idea was to centrally manage policies using a flow-based network and a controller with a focus on network security; that is the reason for *Flow* being in the name *OpenFlow*.

After the initial work by Berkeley and Stanford, companies such as Nicira and Big Switch Networks started to raise significant amounts of venture capital funding to help push their products with ideas on a flow-based controlled network, but at that time no standards were yet published. A protocol was needed to move network control out of proprietary network switches and into control software that was open source and locally managed. This is the reason that the name *OpenFlow* has the word *Open* in it.

By 2011, the **Open Networking Foundation (ONF)** had been created with the aim of standardizing emerging technologies for networking and data center management. The founding members were Google, Facebook, and Microsoft, while Citrix, Cisco, Dell, HP, F5 Networks, IBM, NEC, Huawei, Juniper Networks, Oracle, and VMware joined later.

The ONF working group released the first version of the OpenFlow protocol in December 2009, and in February 2011 they made version 1.1 public. The most updated version is from March 2015 – version 1.5.1 (`https://opennetworking.org/wp-content/uploads/2014/10/openflow-switch-v1.5.1.pdf`).

SDN architecture

The simple architecture of an SDN is shown in *Figure 2.4*. The SDN controller has **northbound interfaces** (**NBIs**) toward business-level applications and **southbound interfaces** (**SBIs**) toward network devices.

To communicate with network devices, the SBI requires a control protocol. It is desirable for the control protocol to be OpenFlow; however, other protocols can be used if the device does not support it, such as Cisco OpFlex, SNMP, or even CLI via SSH (this will be covered in the next chapter).

The NBI is used to collect information from the business or for the business to collect information from the network (in *Figure 2.4*, this is represented by the application plane), for instance, allowing administrators to access the SDN controller to retrieve information about the network. Access to the controller is normally done via an API protocol.

Normally, the NBI is used for the following:

- Getting information from devices
- Getting the status of physical interfaces
- Configuring devices
- Constructing data flows between devices

But the available methods on the NBI API will depend on the SDN application and what the vendor made available.

> **Important note**
> It's important to emphasize that the NBI API for the SDN has no responsibility for managing the network devices, such as attributing configuration or doing software updates. The main responsibility of the SDN NBI API is to allow administrators and businesses to give directions to the SDN controller in order to make decisions on how traffic will flow through the network devices based on pre-defined criteria.

Now, let's look at the simple architecture of an SDN:

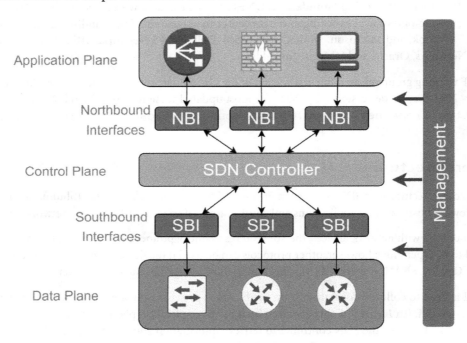

Figure 2.4 – Basic SDN architecture

Despite being used in SDN and being a very well-known term in the internet community, OpenFlow's future might be not that bright. Let's find out why.

OpenFlow and its future

Looking at how the OpenFlow standard track is being updated and how vendors are implementing it, its future doesn't look promising.

The first usable version of OpenFlow was published in 2011, known as version 1.1. Since then, updates have been incorporated until 2015 with version 1.5.1. But more than six years have passed and no updates have been published yet.

Version 1.6 of OpenFlow has been available since 2016, but only for members of the ONF, which does not help the user's confidence in OpenFlow's future.

In addition to the lack of updates, Cisco (one of the major network vendors) has been working on its own version of OpenFlow called OpFlex since 2014 because it saw limitations in OpenFlow's approach. Cisco also has made OpFlex open, allowing others to use without restriction and has started working on an RFC to publish OpFlex (`https://datatracker.ietf.org/doc/draft-smith-opflex/`).

So, the SBIs described in *Figure 2.4* do not necessarily use OpenFlow. Today, SDN implementations vary and may use different types of SBIs that are associated to the methods available for device communication for creation of the traffic flow policies.

Other methods and protocols besides OpenFlow are being used with SDN communication, such as OpenStack, OpFlex, CLI vis SSH, SNMP, and NETCONF, among others.

As we've seen in this section, the SDN is a very well-delineated concept on how to work with programmable networks; however, because of the lack of OpenFlow adoption, SDNs have become more of a concept than a standard. From now on, you should have enough knowledge to decide whether your network automation should follow OpenFlow or not.

Understanding cloud computing

The goal of **cloud computing** is to allow users to benefit from virtual technologies without having in-depth knowledge about them. The objective of cloud computing is to cut costs and help users focus on their core business instead of the physical infrastructure.

Cloud computing advocates for **Everything as a Service (EaaS)**, including **Infrastructure as a Service (IaaS)**, which is implemented by providing high-level APIs used to abstract various low-level details of underlying network infrastructure such as physical computing resources, locations, data partitioning, scaling, security, and backup.

Our focus here will be the networking services offered by cloud computing, which we usually refer to as cloud networking services.

Commercial cloud computing

Perhaps the most popular cloud computing service today is the 2002 Amazon-created subsidiary called **Amazon Web Services (AWS)**. AWS uses its proprietary API to offer cloud services; one of them is created by using **AWS CloudFormation** to provide infrastructure as code.

In 2008, Google started offering cloud services; in 2010, Microsoft started offering Microsoft Azure; in 2011, IBM announced IBM SmartCloud; and in 2012, Oracle start offering Oracle Cloud.

There are hundreds of other providers and a list of all can be found at the following link: `https://www.intricately.com/industry/cloud-hosting`.

The OpenStack Foundation

The **OpenStack Foundation** was the first initiative created by NASA and Rackspace to start an open source cloud service software. The foundation eventually changed its name to the **OpenInfra Foundation**, and today they have more than 500 members. Their work has been tremendous, and they created a great set of open source code for cloud computing. More details can be found at `https://openinfra.dev/about/`.

Cloud Native Computing Foundation

It sounds a bit confusing, but the **CloudStack Foundation** and the **Cloud Native Computing Foundation** (**CNCF**) focus on different aspects of cloud services. The CNCF was basically created by Kubernetes as a Linux-container-based idea, and CloudStack is a bit older and based on VMs.

The CNCF is a Linux Foundation project that was founded in 2015 to help advance Linux container technologies and help to align the tech industry around its evolution. It was announced alongside Kubernetes 1.0, which was given to the Linux Foundation by Google as a seed technology.

We've covered quite a bit about cloud computing, but the key takeaway is that even though it was originally intended to add programmability to computers, cloud computing is also growing in the network space. One of the most programmable networks in this space is OpenStack, which we are going to explore next.

Using OpenStack for networking

In contrast to OpenFlow, OpenStack has been busy and promising. It started in 2010 after a joint project between NASA and Rackspace. Rackspace wanted to rewrite the infrastructure code running its cloud servers and at the same time, Anso Labs (contracting for NASA) had published beta code for Nova, a Python-based cloud computing fabric controller.

By 2012, the OpenStack Foundation was established to promote OpenStack software to the cloud community. By 2018, more than 500 companies had joined the OpenStack Foundation. By the end of 2020, the foundation announced that would change its name starting in 2021 to the **Open Infrastructure Foundation**. The reason is that the foundation started to add other projects to OpenStack, and therefore the name would not reflect their goals.

OpenStack tracks its versions with different names; the first version in 2010 was called Austin, which included two components (Nova and Swift). By 2015, the new version of OpenStack had arrived, which was called Kilo and had 12 components. By October 2021, OpenStack Xena had been released, with 38 service components (`https://docs.openstack.org/xena/`).

For us, what matters in OpenStack are the components that will allow us to automate the network infrastructure. Although not designed for physical devices, the API methods for networking might be extended to physical devices instead of being only used in cloud virtual environments.

OpenStack Neutron

The goal of OpenStack is to create standard services that allow software engineers to integrate their applications with cloud computing services. The Xena version released in October 2021 had 38 services available.

One of the most important services for networking is called **Neutron** (or **OpenStack Networking**), which is an OpenStack project aimed at providing *network connectivity as a service* between interface devices. It implements the OpenStack networking API.

> **Important note**
> Neutron API definitions can be found at the following link: `https://docs.openstack.org/api-ref/network/`.

Neutron manages all networking configurations for the **virtual networking infrastructure** (**VNI**) and the access layer aspects of the **physical networking infrastructure** (**PNI**). It also enables projects to create advanced virtual network topologies, which may include services such as a firewall and a VPN. It provides networks, subnets, and routers as object abstractions. Each abstraction has functionality that mimics its physical counterpart: networks contain subnets, and routers route traffic between different subnets and networks.

For more details on Neutron, visit the following link: `https://docs.openstack.org/neutron`.

The Neutron API

The **Neutron API** is a RESTful HTTP service that uses all aspects of the HTTP protocol, including methods, URIs, media types, response codes, and more. API clients can use existing features of the protocol, including caching, persistent connections, and content compression.

As an example, let's look at the `HTTP GET` BGP peers' method.

To obtain a list of BGP peers use a `HTTP GET` request to `/v2.0/bgp-peers`. The possible responses are as follows:

- Normal response codes: `200`
- Error response codes: `400, 401, 403`

Fields that can be added to the API request:

Name	In	Type	Description
fields (optional)	Query	String	The fields that you want the server to return. If a `fields` query parameter is not specified, the networking API returns all attributes allowed by the policy settings. By using the `fields` parameter, the API returns only the requested set of attributes. A `fields` parameter can be specified multiple times. For example, if you specify `fields=id&fields=name` in the request URL, only the `id` and `name` attributes will be returned.

Table 2.1 – API request fields

The parameters that are returned in the API response are as follows:

Name	In	Type	Description
bgp_peers	Body	Array	A list of `bgp_peer` objects. Each `bgp_peer` object represents real BGP infrastructure, such as routers, route reflectors, and route servers.
remote_as	Body	String	The remote autonomous system number of the BGP peer.
name	Body	String	A more descriptive name of the BGP peer.
peer_ip	Body	String	The IP address of the peer.
id	Body	String	The ID of the BGP peer.
tenant_id	Body	String	The ID of the tenant.
project_id	Body	String	The ID of the project.

Table 2.2 – API response fields

The following is an example of the API response:

```
{
    "bgp_peers":[
    {
        "auth_type":"none",
        "remote_as":1001,
        "name":"bgp-peer",
        "tenant_id":"34a6e17a48cf414ebc890367bf42266b",
```

```
            "peer_ip":"10.0.0.3",
            "id":"a7193581-a31c-4ea5-8218-b3052758461f"
    }
    ]
}
```

The API is well documented at the following link: `https://docs.openstack.org/api-ref/network/`.

As we have seen in this section, OpenStack is perhaps the cloud computing platform that is closest to network programming, as demonstrated by the CloudStack Neutron API. Additional features are probably going to be added as more network elements are migrated to the cloud. You should now be familiar with OpenStack terms and be able to explore them in depth if necessary.

Summary

In this chapter, we talked about how programmable networks have evolved up until the present day. We discussed the history of data plane and control plane separation. We've seen how network virtualization has improved over time. We also looked at some technologies and standards for SDNs and cloud networking, such as OpenFlow and OpenStack.

You now have the knowledge required to understand why some technologies are used today to automate and code networks.

In the next chapter, we're going to dive deeper into the methods, protocols, and standards used to configure and communicate with network devices.

3
Accessing the Network

In the previous chapter, we looked into programmable networks and their history. One of the ideas we explored was **Software-Defined Networking (SDN)**, where we saw why the separation between the data plane and the control plane was important. One important point of SDN was its architecture and how it separates the **North Bound Interface (NBI)** and **South Bound Interface (SBI)**. In this chapter, we are going to explore how to access the network devices that can be interpreted as the SBI for SDN when OpenFlow is not available on the device.

As we have seen before, OpenFlow is not a widely adopted protocol, and its availability is limited to a few manufacturers and devices. Therefore, if you are planning to use SDN, you may need to use the available native methods to configure the devices.

Network access is not only used for SDN but also for a variety of software, such as network configuration, configuration audit, upgrade tools, and automation, among others. Additionally, devices usually have multiple methods or protocols, some of which may be better than others.

In this chapter, we will explore the most common methods and protocols for accessing network devices for our network automation. Since devices have multiple methods, we will aim to give you enough information so that you can choose the one that is most appropriate for your network automation code.

We are going to explore the following topics:

- Working with the CLI
- Using SNMP
- Employing NETCONF
- Adopting gRPC
- Operating with gNMI

Working with the CLI

The **Command-Line Interface** (**CLI**) is perhaps the most widely available method for accessing a network device. It is a term imported from computers, which was a replacement for **teletypewriter** (**TTY**) machines. A CLI is normally implemented by using a program that runs inside the device to interpret the keys being typed. Early implementations of the CLI program monitored the device's serial port, where a terminal with a keyboard was connected to communicate.

In UNIX, the CLI program was called a shell, and the first shell, called the **V6 shell**, was created in 1971 by Ken Thompson at Bell Labs. The **Bourne shell** was introduced in 1977 as a replacement for the V6 shell. Although the UNIX shell is used as an interactive command interpreter, it was also intended to be a scripting language and contains most of the features that are commonly considered to produce structured programs.

Network devices use a simplified version of a shell for their CLIs. Let's explore a bit more what CLI access can offer.

The command prompt

When using the CLI in our network automation work, the command prompt is the most important piece in our code to be interpreted. We will see that the device uses the command prompt to indicate when it is ready to receive a new command.

A **command prompt** (or prompt) is a sequence of characters used in the CLI to indicate readiness to accept a new command, which means it *prompts* the user to take action. A prompt usually ends with one of the following characters: $, %, #, :, >, or -. It also can include additional information, such as the current time, working directory, username, or hostname. On many network devices, the prompt normally ends with $ or %, and for the privileged CLI access mode, it normally ends with #, which is similar to the UNIX superuser, root.

Most of the prompts can be modified by the user – however, the most common information presented in the prompts of network devices is the hostname and sometimes, the username used for login.

The example in *Figure 3.1* shows an **FRRouting** command prompt with the hostname as core-router and the > character at the end, meaning it is waiting for the commands to be placed where the cursor is:

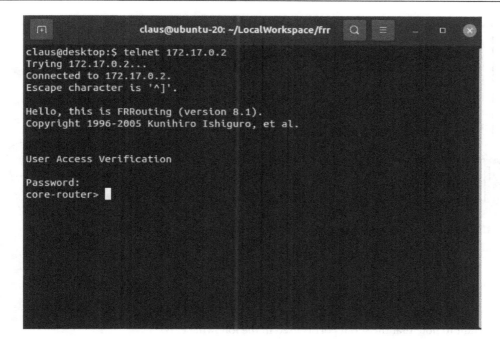

```
claus@desktop:$ telnet 172.17.0.2
Trying 172.17.0.2...
Connected to 172.17.0.2.
Escape character is '^]'.

Hello, this is FRRouting (version 8.1).
Copyright 1996-2005 Kunihiro Ishiguro, et al.


User Access Verification

Password:
core-router>
```

Figure 3.1 – A FRR router command prompt example

Serial access

A network device normally has a special port called a *console* or *serial console*. This port is normally configured to operate at a slow speed, and the common configuration sets the baud rate to 9,600, with some devices able to accept up to 115,200 bits per second. The serial port is normally a **DB9** connector or an **RJ45**, with **RS232** technical specifications. *Figure 3.2* shows an example of the pins used for DB9 and RJ45 serial console connectors.

The program that handles the serial port is normally independent of the device's operating system, allowing this type of port to be used in catastrophic scenarios:

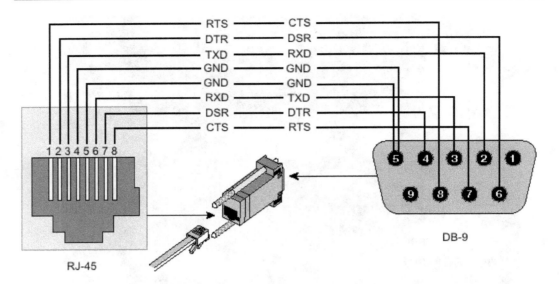

Figure 3.2 – An example of a device's serial port pin configuration

It's important to note that the serial port is used in the following cases:

- Extreme cases when the network device is unresponsible
- For local maintenance in the presence of a technician
- Critical upgrades with the risk of crashes or operation system deletion
- Operation system or hardware failure

The network automation should avoid using the serial port to configure the device because no parallelism exists (only one port) and its speed is limited.

Remote insecure access

For the reasons explained previously, the serial port is slow and can't be used in parallel. Therefore, the best way to access the network device is via remote access, which can be done via **Out-Of-Band (OOB)** or in-band management (described in *Chapter 1*.

How to determine whether the access is secure or not will depend on the protocol used and what kind of network is used to transport this remote access. If access is via an OOB network, it is normally secured and has separated and isolated infrastructure, but if it is in-band, some extra care needs to be taken to avoid some common security breaches.

Insecure applications and protocols

The following applications are normally not encrypted and are easy to eavesdrop on, proxy, or hijack.

Telnet

Telnet is an application that uses TCP port 23 to access a remote device. The data is not encrypted and there is no authentication on the connection. TCP hijacking and eavesdropping are the most common security problems when using Telnet. To make sure these threats are not present, the network path carrying the access must be secured or isolated.

Telnet normally requires a password and sometimes a username as well.

RSH

Remote Shell (RSH) is rarely used. In 1980, it was developed as an alternative to Telnet to provide non-interactive and fast remote command execution. RSH uses TCP on port 514 and does not provide encryption or a password. As IP addresses are used to authenticate access, this protocol is extremely insecure and susceptible to IP spoofing attacks.

If it were not for its lack of security, RSH would be a fast and easy option for running commands and creating network automation.

Remote secure access

To make sure remote access is secure, the data has to be encrypted and the hosts need to have some sort of identity-based authentication to make sure whoever is connecting is allowed to connect.

The most used application for remote CLI access is **Secure Shell (SSH)**. SSH is implemented using the transport protocol TCP with port 22 as the default.

Identity-based authentication

The identity-based authentication on SSH is based on a key fingerprint using SHA-256 (an encryption algorithm). When SSH is used for the first time, SSH asks you to confirm the fingerprint key to make sure the host you are connecting to is the correct one. An example of this fingerprint is shown in *Figure 3.3*:

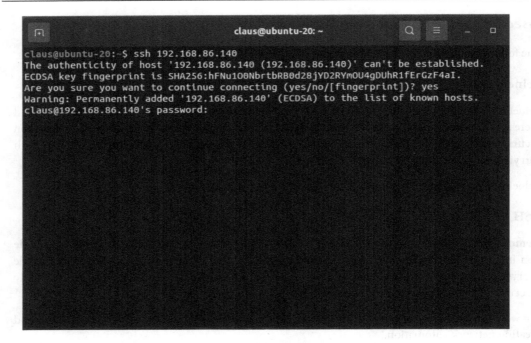

Figure 3.3 – An example of an SSH fingerprint acceptance request

Although SSH is considered a secure application protocol for accessing devices, the identity-based key fingerprint needs to be managed properly to avoid the most common attack, called *man-in-the-middle*. This attack redirects the traffic to another device, pretends it is the end device, and uses this procedure to capture the password of the final target network device, as shown in *Figure 3.4*:

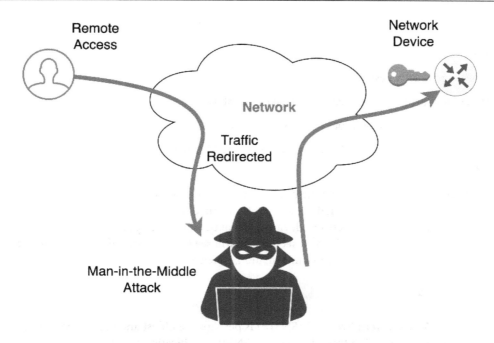

Figure 3.4 – An example of a man-in-the-middle attack on a network device

There are several ways to avoid a man-in-the-middle attack: one is to make sure the network device has an ACL (discussed in *Chapter 1*) that filters incoming IPs that are not part of the management. Another is to make sure identity key management is shared with the agent who is requesting remote access.

Why is this important? Because network automation needs to make sure it uses a secure channel to modify the network and the key management is a shared responsibility between the network operators and the network automation teams.

Here are some pros and cons of using a CLI for network automation:

Pros:

- Available on almost all network devices
- Able to access the whole network device's operating system
- Critical and privileged access
- Remote or local access via serial access without a network

Cons:

- Slow

- Limited parallel access

- When not correctly used, may allow hackers to interfere

- Interactive, requiring prompt

- Data information is not structured (such as JSON or XML), which makes it prone to interpretation errors

To wrap up this section, we've covered the usage of the CLI, which is the oldest interface known for gathering information from network devices. We also learned that a CLI is the only way to collect or configure the device for some network devices. One of the important points is that CLI is always present on the network device. In the next section, we will see how we can improve the interface of the network device for gathering network information, called SNMP.

Using SNMP

Simple Network Management Protocol (**SNMP**) is perhaps the oldest and most used protocol to gather management information from devices. The SNMP specification was first published in 1998 in RFC 2261 and was designed to be simple and fast.

SNMP agents and managers

SNMP defines two identities:

- The manager (or server)

- The agent

The agent is normally a network device and the manager is a network management system.

SNMP uses UDP and TCP as transport protocols with ports 161 and 162. The UDP port 161 is used to gather or set information in an on-demand manner, where the manager sends the request to the agent. The UDP port 162 is used asynchronously without a request from the manager. The agent sends UDP packets to the server whenever necessary. This method is called an SNMP trap and is used to send unsolicited messages, such as alarms or threshold-level breaches.

An SNMP MIB

As described in *Chapter 1*, a **Management Information Base** (**MIB**) is used as an identifier to access the network information variable. The identifier is known as an **Object Identifier** (**OID**), as seen in *Figure 1.8*.

SNMP versions

SNMP comes with version 1, version 2c, and version 3. They were published at different times, and the differences are the available methods, transport protocol, MIB variables, and cryptography. SNMP agents are compatible backward – therefore, an agent that supports version 3 can work with all versions. The following sections present a summary of each version.

SNMPv1

This is the first version and contains most of the MIB variables and methods. It is based on UDP, using a community string as authentication. It only supported 32-bit MIB counters, which is a problem with fast interfaces, as the counter expires quickly.

SNMPv2c

Introduced 64-bit MIB variable counters and the `InformRequest` and `GetBulkRequest` methods. Version 2 was not adopted because of security complexity. Version 2c was then published with simplifications and was widely accepted.

SNMPv3

Introduced cryptography for authentication and privacy. The `Report` method was added.

SNMP primitive methods

SNMP essentially uses the following primitive methods.

- GET: Methods to gather information from the agent:

  - GetRequest: Given an OID, returns the variable associated with it

  - GetNextRequest: Given an OID, returns the next OID with the variable associated with it (used on an SNMP MIB walk function)

  - GetBulkRequest: Given an OID base, returns all OIDs and variables under this OID branch (sometimes causes long responses that can't be interrupted)

- SET: Methods to set a value on the agent:

  - SetRequest: Given an OID and a value, sets it on the MIB

- Response: All responses for SET and GET

- TRAP: Asynchronous information sent from the agent to the manager with an OID and a variable

- InformRequest: Used to send asynchronous information with acknowledgment

SNMP security issues

Because of a lack of cryptography and authentication, SNMPv1 and v2 are vulnerable to IP spoofing attacks, which allow hackers to potentially send `SET` requests to agents compromising the network. Historically, because of this security issue, SNMP is not used to write configuration and only gathers configuration using the `GET` method or `TRAP`.

Here are some pros and cons of using SNMP for network automation:

Pros:

- Easy to implement
- Fast
- Parallelism is easy
- No privileged access

Cons:

- Requires pooling to gather information frequently
- Not normally used for writing information
- Very limited scope of data coverage compared to a CLI
- Security issues when writing

As we saw in this section, SNMP is the oldest and most robust protocol for network management. Although it has security and scope problems for writing configurations, its protocol is light, fast, and easy to read. The following section will cover the protocol developed by the IETF working group to fill the gap in network management concerning configuration. It is called NETCONF.

Employing NETCONF

The **Network Configuration Protocol** (**NETCONF**) is a network management protocol developed and standardized by the IETF in 2006. It provides mechanisms to install, manipulate, and delete the configuration of network devices.

NETCONF operations are implemented on top of a **Remote Procedure Call** (**RPC**) layer. The NETCONF protocol uses **Extensible Markup Language** (**XML**)-based data encoding for the configuration data as well as the protocol messages. The protocol messages can also be exchanged on top of a secure transport protocol such as SSH (RFC 6242) or using TLS (RFC 7589).

Motivation

Up until the early part of the 21st century, the only management protocol available from IETF was SNMP, which was developed in the late 1980s. It became clear that despite what was originally intended, SNMP was not being used to configure network equipment and was mainly being used for gathering network device information (as we have seen previously). The reasons are various, but mainly because SNMP was insecure and had a limited scope compared to a CLI for instance.

In June 2002, the network management community and the Internet Architecture Board got together with network key operators to discuss the real situation on network management protocols and usage. The results of this meeting are documented in RFC 3535 (`https://datatracker.ietf.org/doc/html/rfc3535`).

It turned out that network operators, instead of using SNMP, were primarily using different proprietary CLIs to configure their network devices. The reasons were various, including security issues and the lack of scope to configure or write configs because SNMP was too rigid to do so.

On the other hand, around this time, Juniper Networks had started to use an XML-based network management approach, which was seen by IETF and the network operator community as an opportunity to combine efforts. This led to the creation of the NETCONF working group in May 2003.

In December 2006, with a lot of help from Juniper Networks, the first version of the base NETCONF protocol was published, RFC 4741 (`https://datatracker.ietf.org/doc/html/rfc4741`). After that, several extensions were published in subsequent years (RFC 5277, RFC 5717, RFC 6243, RFC 6470, and RFC 6536, among others). The last revised version of NETCONF is documented in RFC 6241, published in June 2011 (updated by RFC 7803 and RFC 8526).

OpenConfig

OpenConfig is an informal working group of network operators sharing the goal of moving our networks toward a more dynamic, programmable infrastructure by adopting SDN principles such as declarative configuration and model-driven management and operations.

Our initial focus in OpenConfig is on compiling a consistent set of vendor-neutral data models – written in **Yet Another Next Generation** (**YANG**) – based on the actual operational needs from use cases and requirements from multiple network operators.

YANG

YANG is a data modeling language that is used by the NETCONF protocol. YANG can be used to model both configuration data and state data from network devices. It is a modular language representing data structures in the XML format but can also be represented by other formats.

For each network device feature, at least one RFC describes the data model with YANG – for instance, VRRP (in *Chapter 2*) describes the YANG data model in RFC 8347 (`https://datatracker.`

ietf.org/doc/html/rfc8347). Another effort to cover network ACLs (see *Chapter 1*) describes the YANG models in RFC 8519 (https://datatracker.ietf.org/doc/html/rfc8519).

Let's examine the characteristics of YANG and the details of the model a bit more closely.

The data modeling process is hard

It is important to understand that creating a YANG data model for a router function is not an easy task because it has to accommodate possible scenarios from the existing methods of all possible devices. So, it is not a fresh start from scratch, but the task of modeling functions that are already in use in several vendors and devices. Let's take one example – the YANG data model for **Routing Policy**. As you can see on the timeline shown in *Figure 3.5*, the work started in 2015, and after more than 30 drafts, the standard was finally published in October 2021, which meant it took almost 7 years:

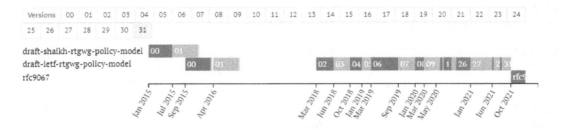

Figure 3.5 – Timeline for creating a YANG data model for Routing Policy

It would be easier if each vendor had its own YANG data model, but then that would remove the general dependency.

NETCONF

NETCONF uses client-server communication based on RPCs. With NETCONF, server configurations are stored in a NETCONF configuration datastore that follows a YANG data format specification. To change or update data, a client sends an XML-based remote procedure call over one of the secure transfer methods, and the server replies with XML-encoded data.

NETCONF has four layers, as shown in *Figure 3.6*, extracted from the original RFC 6241:

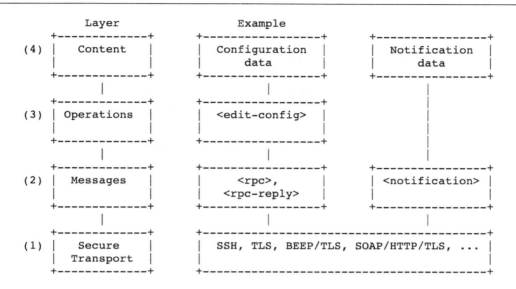

Figure 3.6 – The four layers described in the RFC 6241

Let's summarize each of the layers:

- **Content** layer: Consists of configuration data and notification data. Valid content is defined in the YANG specification.

- **Operations** layer: This layer is also defined by YANG. It details the type of action or query for a given NETCONF transaction. For example, you can use NETCONF to edit, delete, or copy configuration data. Valid operations are get, get-config, edit-config, copy-config, delete-config, lock, unlock, close-session, and kill-session.

- **Messages** layer: Provides a mechanism for encoding **remote procedure calls** (**RPCs**). A message may be a <rpc> request from a client or a <rpc-reply> from a server. RFC6241 also added notifications to this layer.

- **Secure Transport** layer: This layer deals with the protocols used to transmit NETCONF messages. SSH, TLS, and HTTP are a few of the protocols associated with this layer.

RESTCONF

The RESTCONF protocol is a proposed standard defined in RFC 8040 (`https://www.rfc-editor.org/rfc/rfc8040.html`). NETCONF and RESTCONF are similar in terms of their capabilities, but RESTCONF came later in 2017 with a **Representational State Transfer (REST)ful API** model using HTTP. They both allow administrators to query information or modify settings using a client-server model. RESTCONF is different in a few key ways:

- RESTCONF uses HTTP

- RESTCONF supports both JSON and XML

- RESTCONF does not have the concept of transaction and therefore does not have the `lock` concept as NETCONF does

RESTCONF is not intended to replace NETCONF. Rather, it was created to allow the use of a RESTful API that can be used to query and configure devices with NETCONF or YANG configuration datastores.

Figure 3.7 shows a table extracted from RFC 8040 that demonstrates the overlap between RESTCONF and NETCONF calls:

```
+----------+-----------------------------------------------------------+
| RESTCONF | NETCONF                                                   |
+----------+-----------------------------------------------------------+
| OPTIONS  | none                                                      |
|          |                                                           |
| HEAD     | <get-config>, <get>                                       |
|          |                                                           |
| GET      | <get-config>, <get>                                       |
|          |                                                           |
| POST     | <edit-config> (nc:operation="create")                     |
|          |                                                           |
| POST     | invoke an RPC operation                                   |
|          |                                                           |
| PUT      | <copy-config> (PUT on datastore)                          |
|          |                                                           |
| PUT      | <edit-config> (nc:operation="create/replace")             |
|          |                                                           |
| PATCH    | <edit-config> (nc:operation depends on PATCH content)     |
|          |                                                           |
| DELETE   | <edit-config> (nc:operation="delete")                     |
+----------+-----------------------------------------------------------+
```

Figure 3.7 – The overlap between RESTCONF and NETCONF methods extracted from RFC 8040

Here are some pros and cons of using NETCONF or RESTCONF:

Pros:

- Incorporate network specification
- IETF standards
- No privileged access
- Allow stream event notifications
- Programmatic device configuration

Cons:

- Not all device capabilities are covered in YANG
- Adoption of NETCONF has been really slow
- NETCONF transport is limited and implementation is old
- Not that efficient

This section summarizes how NETCONF, RESTCONF, and YANG are used to interact with network devices. The transaction states of NETCONF make it a powerful tool for network configuration. Despite its good base of IETF standards, NETCONF is not efficient enough to handle some of the network automation we want, such as collecting data at a high-frequency rate. In the following section, we are going to explore a newer protocol called gRPC.

Adopting gRPC

gRPC was published in 2015 as an open source RPC framework. It is one of the most promising protocols to be used in automation because it is easy to create a program and add methods to obtain or set configuration on the network device.

gRPC does not directly use TCP for transport, but HTTP/2 instead, which was published in 2015 to overcome the limitations of HTTP/1.1. While it is backward compatible with HTTP/1.1, HTTP/2 brings many added advanced capabilities, such as the following:

- **Binary framing layer**: Request and response is divided into small messages and framed in binary format, making message transmission efficient
- **Bidirectional full-duplex streaming**: Here, the client can request and the server can respond simultaneously
- **Flow control (used in HTTP/2)**: Enables the detailed control of memory used for the network buffers

- **Header compression**: Everything in HTTP/2, including headers, is encoded before it is sent, significantly improving performance

- **Asynchronous and synchronous processing**: Can be used to perform different types of interaction and streaming RPCs

All these features of HTTP/2 enable gRPC to use fewer resources, resulting in reduced response times between clients and servers.

To ensure the security of gRPC, TLS end-to-end encryption can be used, and authentication can use SSL or TLS with or without token-based authentication or the need to define your own authentication system by extending the provided code (more on authentication can be found at `https://grpc.io/docs/guides/auth/`).

The letter g

Initially, in version 1.0 of the protocol, the letter *g* was a recursive reference to the name gRPC, but as later versions were published, another word was added, making the name a bit of code entertainment. For instance, in version 1.1, the word was *good*, in version 1.2, it was *green*, and for version 1.42, it was *granola*. A complete list of names used for the letter *g* can be found in the gRPC source code here: `https://grpc.github.io/grpc/core/md_doc_g_stands_for.html`.

Motivation

Google has used a single general-purpose RPC infrastructure called **Stubby** to connect the large number of microservices running within and across Google data centers for more than a decade. That motivated Google to publish and sponsor the creation of gRPC.

Letter from the gRPC team on Monday, October 26, 2015

The gRPC team is excited to announce the immediate availability of gRPC Beta. This release represents a major step forward in API stability, with most API changes in the future being additive in nature. It opens the door for gRPC use in production environments.

We updated grpc.io documentation to reflect the latest changes and released language-specific reference documentation. In the release notes on GitHub for Java, Go, and all other languages, you will find information on what has changed.

We would like to thank everyone who contributed code, gave presentations, adopted the technology, and participated in the community. We look forward to 1.0 with your support!

Overview

gRPC uses the concept of a client and server application. Client applications can directly invoke server applications on remote machines just as if they were local objects. gRPC is based on the idea of defining a service and specifying methods that can be called remotely using their parameters and return types. The server implements this interface and runs a gRPC server to handle client calls. The client has a stub (just referred to as a client in some languages) that provides the same methods as the server.

In the world of network automation, the gRCP client is actually our automation software and the gRPC server is the network device, as illustrated in *Figure 3.8*:

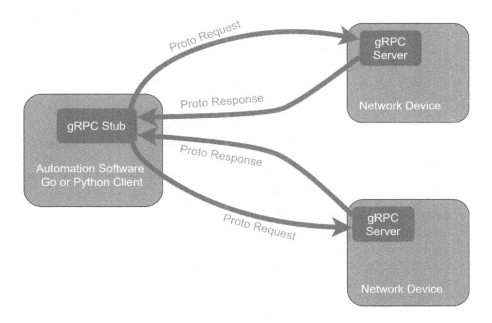

Figure 3.8 – Basic request and response for gRPC

The gRPC server and client are not required to use the same programming language. Today, there are several implementations of different languages, whether Go, Python, Java, or Ruby. A complete list of the languages supported can be found here: https://grpc.io/docs/languages/.

Protobuf

By default, gRPC uses **Protocol Buffers** (**Protobuf**), which is another open source mechanism for serializing data created by Google. Although Protobuf is the default, gRPC can also use JSON instead, but that is less efficient, as we are going to see.

Protobuf is a language- and platform-neutral mechanism for serializing data, like JSON or XML but much smaller, faster, and simpler. The structure of the data is defined once and then a specially generated source code is used to easily write and read this structured data from a variety of data streams with any programming language.

More about Protobuf can be found here: `https://developers.google.com/protocol-buffers`.

gRPC and network telemetry

During our network automation work, we are going to experience a series of limitations, especially in gathering network information in an effective way. So, let's explore the following example.

Imagine a network with 500 devices, with 50 interfaces on each device on average. Each interface needs to gather multiple variables, such as the current state, the error rate, drop counts, packet-in counts, or packet-out counts. If we consider a conservative approach, such as only collecting 10 variables per interface, for this network example, we are going to gather information from 10 variables x 50 interfaces x 500 devices, which adds up to 250,000 variables.

The other point to consider is the frequency of the data. In the 90s, network management required information from the network every 5 minutes and that was sufficient for handling failures and troubleshooting, but nowadays, the interval is much smaller. We are looking to gather information at intervals of less than 1 minute, ideally every 30 or 10 seconds. The reason is that troubleshooting and failure resolution can occur faster when detecting failures quickly.

So, in our example, 250,000 variables every 10 seconds give us an enormous amount of data using traditional polling mechanisms such as SNMP. However, one important point to note here is most of the content of the variables might not change at all, such as the counters for interfaces when there is no traffic, the state of the interface when nothing has changed, the interface discard counter when there aren't any, or the error rate when the interface is perfectly fine. Therefore, several or even the majority of the contents of the network variables are not going to change that often, meaning that pooling mechanisms are inefficient and accumulate redundant information over time. What would be better than polling? Streaming telemetry. Streaming telemetry allows devices to send incremental updates continuously as soon as changes occur. In this way, the collection of network information can be done more effectively than with pooling.

gRPC supports bidirectional streaming, which gives this protocol a huge advantage compared to the others we saw so far for data collection.

Code examples using gRPC

To make the example more realistic for network automation, let's have a service on the routers that can return the following information:

- Return the memory utilization in percent
- Return the CPU utilization in percent
- Return the router uptime in seconds

Our examples will create a client gRPC stub to communicate to the router, which will be the gRPC server, as depicted in *Figure 3.8*. We are only going to demonstrate the client side and we are going to assume the gRPC server on the router has already been implemented.

The Protobuf file

The Protobuf file definition is a single part of the code that is not tied to any language. The same file definition is used on the client and the server. It is compiled once and feeds the client and server programs to interpret the data used to generate the RPCs. For our example, the Protobuf file would look as follows:

```
service RouterStatus {
   rpc GetStatus (StatusRequest) returns (StatusReply);
}
message StatusRequest {}
message StatusReply {
   double memory = 1;
   double cpu = 2;
   int32 uptime = 3;
}
```

An example using Python

Here is an example using Python. The import name, r_grpc, compiles the code for Python from the Protobuf file:

```
import grpc
import routerstatus_pb2
import routerstatus_pb2_grpc as r_grpc
def run():
    address = "router:50051"
    with grpc.insecure_channel(address) as channel:
        stub = r_grpc.StatusStub(channel)
        r = stub.GetStatus(r_grpc.StatusRequest())
        print("Memory:{.2f}% CPU:{.2f}%, Uptime:{d}s\n".
format(r.memory, r.cpu, r.uptime))
if __name__ == '__main__':
    run()
```

An example using Go

Here is an example using a Go program client. Note that pb (used on the import) is the code compiled for the Protobuf:

```go
import (
    "context"
    "log"
    "time"
    "fmt"
    "google.golang.org/grpc"
    pb "example/routerstatus"
)
func main() {
    address = "router:50051"
    // Set up a connection to the server.
    conn, err := grpc.Dial(address, grpc.WithInsecure(), grpc.
WithBlock())
    if err != nil {
        log.Fatalf("did not connect: %v", err)
    }
    c := pb.NewStatusClient(conn)
    ctx, cancel := context.WithTimeout(context.Background(),
time.Second)
    defer cancel()
    r, err := c.GetStatus(ctx, &pb.StatusRequest{})
    if err != nil {
        log.Fatalf("could not get router status: %v", err)
    }
    fmt.Printf("CPU:%.2f%%, Memory:%.2f%%, Uptime:%ds",
r.GetCpu(), r.GetMemory(), r.GetUptime())
}
```

Here are some pros and cons of using gRPC for network automation:

Pros:

- Secure
- Fast
- Parallelism is easy
- No privileged access is possible
- Flexible and can expose any local device command using the gRPC server

Cons:

- Not many network devices have gRPC capability

In this section, we saw that gRPC is a powerful protocol to use for network automation. However, it is not well integrated into network devices yet. The majority of the new operating systems on network devices come with this capability. In the next section, a higher-level protocol called gNMI will be used to make better use of the gRPC protocol for network automation.

Operating with gNMI

As we saw before, gRPC is probably the most appropriate protocol for working with devices in terms of performance. However, it is actually a generic protocol to be used in any client and server interaction – not only network devices but also computer servers. For this reason, **gRPC Network Management Interface (gNMI)** was created.

gNMI is an open source protocol specification created by the OpenConfig working group that is used to communicate to and from network devices using YANG (discussed in the *NETCONF* section). In other words, gNMI was created to utilize the good work done by people defining the network specification data using YANG but with a more modern protocol such as gRPC instead of NETCONF.

Protocol layers

gNMI uses gRPC. For that, it has to translate the YANG data description into Protobuf to serialize the communication, as illustrated in *Figure 3.9*. At the bottom of the diagram is a normal gRPC connection over HTTP/2 and TLS. The gRPC code is auto-generated from the gNMI Protobuf model and gNMI carries the data modeled in YANG, which can support encoding in JSON, like the example below.

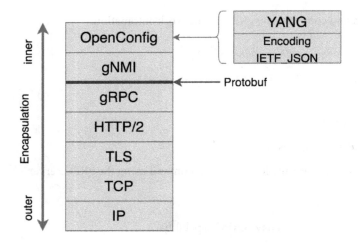

Figure 3.9 – gNMI protocol layers

The data model

gNMI uses a data model called **path** for encoding and decoding. Path encoding in gNMI uses a structured format and is encoded as an ordered list (such as a slice or array) of `PathElem` messages. Each `PathElem` consists of a name encoded as a string. An element's name must be encoded as a **UTF-8** string. Each `PathElem` may optionally specify a set of keys, specified as a `map<string,string>` (dictionary or map).

The root path, `/`, is encoded as a zero-length array (slice) of `PathElem` messages. Here are some example declarations in Go and Python:

- **Go**: `path := []*PathElem{}`
- **Python**: `path = []`

A human-readable path can be formed by concatenating elements of the prefix and path using the `/` separator.

So, let's see the following representation: `/interfaces/interface[name=Ethernet1/2/3]/ state`.

This is specified as follows:

```
<elem: <name: "interfaces">elem: <name: "interface"key: <key:
"name"value: "Ethernet1/2/3">>elem: <name: "state">>
```

The communication model

The communication model uses a target and client as follows:

- **Target**: The device within the gNMI that acts as the owner of the data that is being manipulated or collected. Typically, this is our network device.

- **Client or collector:** The system using the gNMI to query or modify data on the target or act as a collector for streamed data. Typically, this is the network management system or our automation code.

Similar to gRPC, the server actually is on the network device, as depicted in *Figure 3.10*:

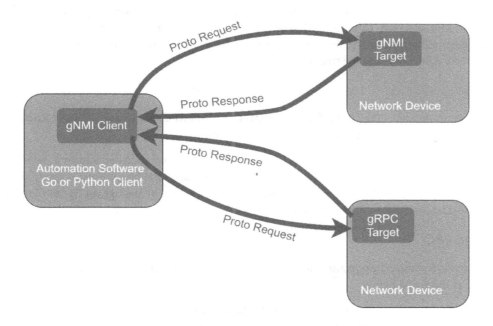

Figure 3.10 – gNMI target and client communication

Service definition

The gNMI service is based on RPC calls called `Capabilities`, `Get`, `Set`, and `Subscribe`, which will be detailed in the following sections.

Capabilities RPC

A client can discover the capabilities of the target using the **Capabilities RPC**. The CapabilityRequest message is sent by the client to interrogate the target. The target must then reply with a CapabilityResponse message that includes its gNMI service version, the versioned data models it supports, and the supported data encodings. This information is used in subsequent RPC messages from the client to indicate the set of models that the client will use for the Get and Subscribe RPC calls and the encoding to be used for the data.

Get RPC

The **Get RPC** provides an interface where a client can request a set of paths to be serialized and transmitted to it by the target. The client sends a GetRequest message to the target, specifying the path that is to be retrieved. Upon reception of the GetRequest message, the target serializes the requested path and returns a GetResponse message. This connection is short-lived and the target closes the Get RPC following the transmission of the GetResponse message.

Set RPC

Modifications to the state of the target are made through the **Set RPC**. A client sends a SetRequest message to the target indicating the modifications it desires.

A target receiving a SetRequest message processes the operations specified within it, which are treated as a transaction. In response to a SetRequest message, the target must respond with a SetResponse message. For each operation specified in the SetRequest message, an UpdateResult message must be included in the response field of SetResponse.

Subscribe RPC

This perhaps is the most important call on the gNMI because it is the one that allows Streaming Telemetry, as discussed before.

When a client wishes to receive updates relating to the state of data instances on a target, it creates a subscription via the **Subscribe RPC**. A subscription consists of one or more paths, with a specified subscription mode. The mode of each subscription determines the triggers for updates to the data sent from the target to the client.

All requests for new subscriptions are encapsulated within a SubscribeRequest message, which itself has a mode describing the longevity of the subscription. A client may create a subscription that has a dedicated stream to return one-off data (ONCE); a subscription that utilizes a stream to periodically request a set of data (POLL); or a long-lived subscription that streams data according to the triggers specified within the individual subscription's mode (STREAM). For Streaming Telemetry, the mode is set to STREAM.

gNMI-gateway

gNMI-gateway is open source software that was initially developed by Netflix and then released as part of the OpenConfig working group to collect and distribute OpenConfig-modeled gNMI data from network devices.

The motivations to create gNMI-gateway were various, as follows:

- First, there were not many open source services available to consume and distribute OpenConfig-modeled gNMI Streaming Telemetry data.

- Second, there was a lack of failure tolerance for the client and target connection using gNMI data streaming, making Streaming Telemetry vulnerable. As the client dies, the streamed data is lost until another subscription takes place.

- The third was the lack of supporting multiple consumers. If multiple departments in a company want data from a network device or a group of network devices, it would be necessary for all of them to send subscriptions to the targets. With clustering functionality and replication in gNMI-gateway, it is possible to avoid unnecessarily duplicating gNMI connections to targets and offer the same data to multiple customers.

- Fourth, there was a lack of unifying gNMI clients with non-gNMI clients. gNMI-gateway allows either gNMI clients or non-gNMI clients to gather information.

Figure 3.11 shows a single instance of gNMI-gateway with gNMI clients and non-gNMI clients, also known as exporters. **Apache Kafta** (`https://kafka.apache.org/`) is one piece of software that can be used as an exporter; another one already implemented is **Prometheus** (`https://prometheus.io/`):

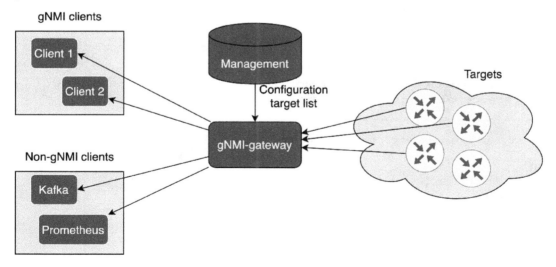

Figure 3.11 – A single instance of gNMI-gateway

The redundancy can be obtained by using multiple instances of gNMI-gateway which is implemented by using **Apache Zookeeper** (`https://zookeeper.apache.org/`), as illustrated in *Figure 3.12*. If only one instance is running, there is no need to use Apache Zookeeper:

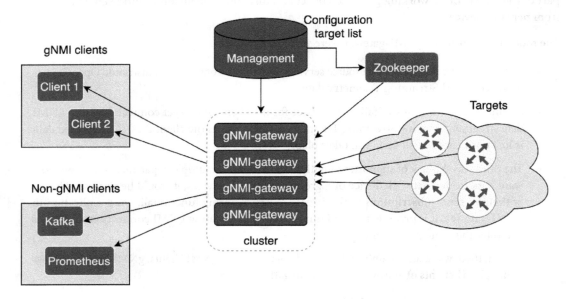

Figure 3.12 – Multiple instances of gNMI-gateway

More on gNMI-gateway can be found here: `https://github.com/openconfig/gnmi`.

For reference, here is a full presentation on gNMI-gateway presented at 2020's NANOG Webcast: `https://nanog.org/news-stories/nanog-tv/nanog-80-webcast/gnmi-gateway/`.

> **Important note**
> Full specification on gNMI can be found here: `https://github.com/openconfig/reference/blob/master/rpc/gnmi/gnmi-specification.md`.

Here are some pros and cons of using gNMI for network automation:

Pros:

- Secure
- Fast
- Parallelism is easy
- No privileged access is possible

- Incorporates YANG specification for networking
- Allows easy Streaming Telemetry
- Capable of adding gNMI-gateway with full redundancy

Cons:

- Not many network devices have gNMI capability

Summary

In this chapter, we covered the major methods used by software engineers to interact with network devices. Not many more methods are available, so I would assume we have covered perhaps 99.9% of all existing methods at the time of writing.

Using the information provided in this chapter, you can choose which method or methods to incorporate into your network automation code. In most cases, you won't be able to cover all scenarios using just one method; you will probably have to combine two or more methods.

The next chapter will explore how files can be used to define a network. We will discuss the pros and cons of each type of file that's available.

4

Working with Network Configurations and Definitions

One important point in network automation is how the configuration is organized and how we can automate our network in a scalable way. In this chapter, we are going to explore how to work with a network configuration and how to define it for effective use with network automation. We want to build scalable and future-proof solutions.

Why do we care about configuration and network definition? Why is it important which file to use? How can we create a lifelong definition? How can we use this to help network automation? Let's explore answers to these questions in this chapter.

We are going to explore the following:

- Describing the configuration problem
- Helping network automation using definitions
- Creating network definitions
- Exploring different file types

Technical requirements

The source code described in this chapter is stored in the GitHub repository at `https://github.com/PacktPublishing/Network-Programming-and-Automation-Essentials/tree/main/Chapter04`.

Describing the configuration problem

Several production networks out there have their configuration applied to network devices without any additional external definition. Some of them have network diagrams describing the network, but the majority have outdated or incomplete diagrams. Therefore, in most cases, you might need to read running device configurations to understand the details of the network operation.

In some network providers, diagrams are used for an initial understanding of the network or as an overall overview. Once the engineers have enough confidence in their network, the diagrams are just ignored or not used anymore. Some would update their diagrams, but for most engineers, this task is not a priority and usually is left behind.

It is also common for some engineers to apply configuration fixes directly into production devices to solve catastrophic or urgent failures. In other cases, additional configuration is applied temporarily for troubleshooting, but never removed. These configuration changes are forgotten in some cases that they were applied and the network runs with a configuration difference that is not perceived until a software update is necessary.

Let's discuss the issues further in the following subsections.

Source of truth

Source of truth is a term used in computer networks to describe where the definition is and what all other systems have to rely on when consulting or using the information to create further definitions. The source of truth can be either a file, a router configuration, a database, a memory space, or a network diagram.

We want to have the source of truth as steadily defined as possible; it should not change over a short period of time and should be used as a reference for any other definition.

The majority of network engineers rely on the router configuration as the source of truth, because routers are frequently updated and have the correct definitions to run the network; however, this does not help our network automation.

Ideally, we want the source of the truth to be in a database or files stored in a secure and future-proof environment that can be ready to be read by any system quickly.

The startup configuration and the running configuration

A network device has two configuration states, the **running configuration** and the **startup configuration**. The running configuration is the current configuration on the device's memory that is being used to operate at that present moment. The startup configuration is used to boot the device from *off* to *on*. The difference is that one is volatile and will be deleted once the device is turned off or loses power supply, and the other is permanent and will always exist independent of whether it has a power supply or not.

The startup configuration is normally stored in a non-volatile memory store such as SSD, a flash drive, or a hard drive.

For our automation, we want to have the same running configuration as the startup configuration. When they differ, it might cause automation issues if not well documented.

Configuration states and history

A good network automation design should aim to have multiple configuration states and history, which will help to determine how the network is operating now, how it should operate in the future, and how it operated in the past.

For deployment and automation, it is very useful to separate the configuration into at least four stages: **Desired**, **Approved**, **Applied**, and **Running**, as shown in *Figure 4.1*:

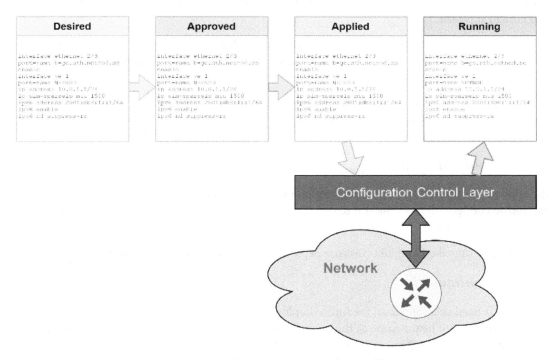

Figure 4.1 – Configuration stages and control layer

We'll discuss each of these in the following sub sections.

Desired-configuration

The **desired-configuration** is used to document the configuration that should be applied in the near future. It is used to check for inconsistencies and to be applied in a test network (or a network simulation) to evaluate future deployments, possible syntax errors, and other functional problems.

Modern networks use the desired-configuration to feed a configuration pipeline that will perform a series of tests, including simulation to validate and anticipate errors. It is also used to test the deployment sequence, and evaluate which devices can be deployed in parallel or which ones need to wait until some have finished being deployed.

Approved-configuration

The **approved-configuration** is the configuration that has passed all approval stages in the configuration pipeline, whether they are automated approval stages, such as syntax checks, or manual stages, done by a human approval.

To enhance confidence, some configuration pipelines will apply the desired-configuration in a network simulation for further function testing, and if everything passes, the configuration pipeline will approve the configuration.

From all configuration stages, the approval-configuration is the one that takes more time to finish as large networks have thousands of routers, and some require sequential functional testing instead of parallel testing. After everything is completed, the configurations are ready to be deployed.

Applied-configuration

The **applied-configuration** is the configuration saved on the device's non-volatile configuration or the startup configuration. At the end of this stage, we confidently know that the configuration has been saved in all routers.

This stage also can take a long time because deployment cannot necessarily be performed in parallel.

Running-configuration

This stage is used as a safeguard for future deployments and approval processes. The **running-configuration** has to be the same as the applied configuration; however, it might not be the same configuration, especially after a catastrophic event that required configuration intervention.

The running-configuration is constantly updated by trigger points such as privileged access to the device, faulty hardware, or any configuration changes.

Using configuration pipelines requires that only the pipeline is able to change the startup configuration of the router. That means network engineers are not able to save configuration changes on the device to avoid this happening.

For catastrophic scenarios where the configuration has to be changed quickly, the configuration pipelines are normally bypassed and the configuration updates are done by manual intervention from network engineers. These changes are normally applied to running-configuration and are not saved to the device. If a configuration audit runs it will show the difference between the running-configuration and the applied-configuration.

Configuration history

It is most desirable to have a history of the configuration for each of the devices for all configuration stages. The configuration history helps us to understand possible failures or improvements by comparing an old setup to a new one. It is also used to build an entire network in simulation for troubleshooting a failure in deployment during configuration updates. A configuration history might also benefit other teams, such as security auditing and capacity planners.

Deployment pipeline

Deployment pipelines are used on large networks that require fast and reliable changes. The pipeline is constructed with modern techniques of simulation and testing, which include extensive use of network automation coding.

An example of a network deployment pipeline can be seen as follows:

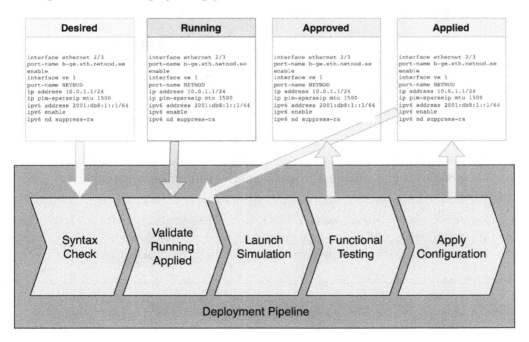

Figure 4.2 – Example of a configuration deployment pipeline

In the preceding figure, we can see how we can use the configuration states in our deployment pipeline. The input to the pipeline is the generated desired configuration, and the approved configuration will only be updated when the final tests have finished. With the use of this workflow, it is possible to automate deployment, reduce mistakes, and allow faster parallel deployments. Let's describe each step of the pipeline process:

1. First, the configuration syntax is checked automatically. If it passes, go to the next steps. If it fails, the pipeline process stops and waits until a valid configuration is presented.

2. Second, the pipeline verifies any differences between the running and applied configurations, which in some cases could be due to an urgent configuration fix. If the configurations are different, it will require manual approval to go to the next steps.

3. Third, the pipeline launches a simulation for the routers in the configuration change scope and deploys the new configuration to the simulated network.

4. Fourth, the pipeline runs functional tests in the simulation that will confirm that the new configuration does not break any functions that are already present. If it passes, it goes to the next step. If it fails, the pipeline stops and awaits rejection or acceptance.

5. Fifth, the pipeline applies the new configuration to the network, following all parallel dependencies and time restrictions.

Network diagrams and automation

Network diagrams are graphical, human-friendly, and readable representations of the network, but they are not easy for machines to read. Most of the time, they are generated by humans from graphical tools such as Visio, Lucid Chart, and Draw.io. When generated by humans, the updates on the diagrams, to reflect the current state, are normally left behind, causing information on the diagram to be outdated.

To have an accurate network diagram, we need to have some sort of diagram generator that reads data from either the router configuration or network definition files.

In our previous example of the deployment pipeline, the diagram generator could read the configuration from any configuration stage and have a different diagram for the running configuration and desired configuration stages.

With automatic diagram generation, an up-to-date diagram can be generated for any stage of the pipeline. These diagrams can be used to help network engineers to troubleshoot a problem or help network designers to understand how to improve the current network functions.

In this section, we discussed how multiple configuration stages can help deploy solutions and how problems can occur when the source of truth is not defined or updated. Next, we will examine how we can create abstract definitions from a device configuration to create a better source of truth.

Using network definitions to aid automation

The previous section explored the configuration stages and how we can rely on them to build a better network deployment pipeline. On the other hand, we have not covered another issue with the router configuration, which is related to router software versions and router vendors.

Using a router configuration as a source of truth has advantages if your network will not update, grow, or change vendors. If your network is not intended to change, you might not need a definition at all. However, as the majority of the network will need to upgrade or grow, it is important to think about getting away from vendor-specific solutions and create vendor-agnostic definitions of your network.

A router vendor has different configuration defaults, which means some configuration lines might not be necessary with one vendor but be required with the other vendor. For network automation, we want to avoid traps like that and have a network source of truth that explicitly says what is necessary to configure. We then have to add a translation layer that will produce a router configuration specifically for that vendor.

Another point is some vendors change default configuration between versions of the same operational system. One version might have extra lines that are not present in the other version. This can also cause problems with our network automation. Again, we want to add a translation layer that is vendor- and version-specific, therefore generating configurations appropriately.

The router configuration render

A router configuration render is a software layer that sits between our agnostic definitions and the desired configuration in our deployment pipeline described earlier. It works like a translator and needs to be aware of the vendor and the version of the router to generate an appropriate configuration. The following figure shows an example of two different configuration renders, one for Juniper and one for Cisco:

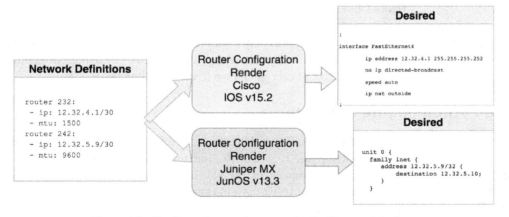

Figure 4.3 – Configuration render examples for Cisco and Juniper

A configuration render can be flexible when generating configuration lines for the same vendor, but it must be aware of the configuration defaults and difference for each vendor and its operating system version. Normally, a render can cover a series of versions and platforms from the same vendor.

Using configuration templates

An easy way to generate a router configuration is to use **configuration templates**. With these, it is easy to construct a generic configuration and then modify by adding key text words to the configuration text that has to be changed.

In our case, the router configuration render will read the definitions from a file and then read a configuration template to generate the router configuration as described in the following diagram. It shows an example for a Juniper router, but it can be used for any other router:

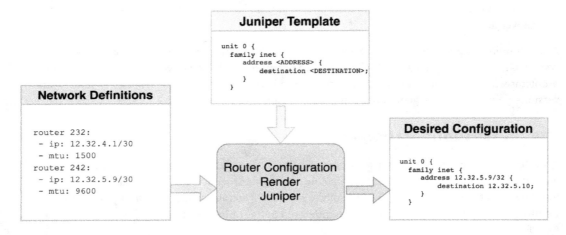

Figure 4.4 – Router configuration render using templates

Creating a template is not difficult and it requires only a sample of the router configuration platform that you want to generate; you simply replace the elements you want with a key string. An example using {{ADDRESS}} and {{DESTINATION}} is shown in the preceding figure.

Using Python engine templates

Jinja is a template engine library for Python that is also commonly referred to as **Jinja2** to reflect the newest release version. Jinja is used to create HTML, XML, and, for us, router configurations. It was created by Armin Ronacher and is open source with the BSD license. Jinja is similar to the Django template engine, but with the advantage of using Python expressions. It uses a text template and, therefore, can be used to generate any text, markup, or even source code, and router configurations in our case.

For router configurations, Jinja is useful because it has a consistent template tag syntax and the router template configuration is extracted as an independent source, so it can be used as a dependency by other code libraries.

The following is an example of an input Jinja template file for a Cisco router (a file called `cisco_template.txt`):

```
hostname {{name}}
!
interface Loopback100
description {{name}} router loopback
ip address 100.100.100.{{id}} 255.255.255.255
!
interface GigabitEthernet1/0
description Connection to {{to_name}} router G0/1
!
interface GigabitEthernet1/0.1{{id}}
description Access to {{to_name}}
encapsulation dot1Q 1{{id}}
ip address 100.0.1{{id}}.2 255.255.255.252
ip ospf network point-to-point
ip ospf cost 100
!
router ospf 100
router-id 100.100.100.{{id}}
network 100.0.0.0 0.255.255.255 area 0
!
```

The following is the input definition file used to feed the configuration render (a file called `router_definitions.yaml`):

```
- id: 11
  name: Sydney
  to_name: Melbourne
- id: 12
  name: Brisbane
  to_name: Melbourne
- id: 13
```

```
name: Adelaide
to_name: Melbourne
```

The following is the Python code used to generate the router configurations:

```python
from jinja2 import Environment, FileSystemLoader
import yaml
env = Environment(loader=FileSystemLoader('templates'))
template = env.get_template('cisco_template_python.txt')
with open('router_definitions.yaml') as f:
    routers = yaml.safe_load(f)
for router in routers:
    router_conf = router['name'] + '_router_config.txt'
    with open(router_conf, 'w') as f:
        f.write(template.render(router))
```

After running the preceding Python script, it will generate three files with three different configurations for Sydney, Brisbane, and Adelaide.

Here is the content of the `Sydney_router_config.txt` output file:

```
hostname Sydney
!
interface Loopback100
description Sydney router loopback
ip address 100.100.100.11 255.255.255.255
!
interface GigabitEthernet1/0
description Connection to Melbourne router G0/1
!
interface GigabitEthernet1/0.111
description Access to Melbourne
encapsulation dot1Q 111
ip address 100.0.111.2 255.255.255.252
ip ospf network point-to-point
ip ospf cost 100
!
router ospf 100
router-id 100.100.100.11
```

```
network 100.0.0.0 0.255.255.255 area 0
!
```

The following is the content of the `Brisbane_router_config.txt` output file, just to show the slight differences between the files:

```
hostname Brisbane
!
interface Loopback100
description Brisbane router loopback
ip address 100.100.100.12 255.255.255.255
!
interface GigabitEthernet1/0
description Connection to Melbourne router G0/1
!
interface GigabitEthernet1/0.112
description Access to Melbourne
encapsulation dot1Q 112
ip address 100.0.112.2 255.255.255.252
ip ospf network point-to-point
ip ospf cost 100
!
router ospf 100
router-id 100.100.100.12
network 100.0.0.0 0.255.255.255 area 0
!
```

Like Sydney and Brisbane, the Adelaide file will be created with the required fields changed. The preceding example is quite simple and has only three keys to be modified: `{{id}}`, `{{name}}`, and `{{to_name}}`. More complex examples can be found in the Jinja documentation at `https://jinja.palletsprojects.com/`.

Using Go engine templates

Since Jinja is limited to just Python, **Go** has a native text template engine that can be used to generate router configurations.

Different from Python, Go templates are executed by applying a data structure to the template text. The template has annotations that refer to the data structure, which is normally a field of `Struct` or `Map` in Go.

For our example using Go, let's use a similar template for the Cisco configuration used for Jinja in the previous example, but with small changes to accommodate Go standards. Let's use the file named cisco_template_go.txt:

```
hostname {{.Name}}
!
interface Loopback100
description {{.Name}} router loopback
ip address 100.100.100.{{.Id}} 255.255.255.255
!
interface GigabitEthernet1/0
description Connection to {{.Toname}} router G0/1
!
interface GigabitEthernet1/0.1{{.Id}}
description Access to {{.Toname}}
encapsulation dot1Q 1{{.Id}}
ip address 100.0.1{{.Id}}.2 255.255.255.252
ip ospf network point-to-point
ip ospf cost 100
!
router ospf 100
router-id 100.100.100.{{.Id}}
network 100.0.0.0 0.255.255.255 area 0
!
```

And for the router definition, the router_definitions.yaml file is the same as that used in the Python example.

The following is the Go code used to generate the same router configurations created in Python:

```
package main
import (
    "io/ioutil"
    "os"
    "text/template"
    "gopkg.in/yaml.v3"
```

```go
)
type Router struct {
    Id      int    `yaml:"id"`
    Name    string `yaml:"name"`
    Toname  string `yaml:"to_name"`
}
type RouterList []Router
func check(e error) {
    if e != nil {
        panic(e)
    }
}
func main() {
    var routers RouterList
    yamlFile, err := ioutil.ReadFile("router_definitions.yaml")
    check(err)
    err = yaml.Unmarshal(yamlFile, &routers)
    check(err)
    templateFile, err := ioutil.ReadFile("cisco_template_
go.txt")
    check(err)
    for _, router := range routers {
        outFile, err := os.Create(router.Name + "_router_
config.txt")
        check(err)
        tmpl, err := template.New("render").
Parse(string(templateFile))
        check(err)
        err = tmpl.Execute(outFile, router)
        check(err)
    }
}
```

The code in Go is similar to Python, but note that in Go, you have to explicitly describe all the fields you are going to read from the `router_definitions.yaml` file. This is done in the code with the `Router` and `RouterList` types (lines 11 and 17 in the preceding code).

In this section, we explored how we can improve network automation by having router configuration renders. We also explored some very useful libraries in Python and Go to be used for router configuration renders. Next, we are going to explore the nuances of creating network definitions.

Creating network definitions

We have seen so far how template engines are extremely useful to create router configurations using network automation. It is also important to define a good router definition, so we can have templates that are less specific, allowing the router definition file to determine how the router should be configured. Additionally, if router definitions are created properly, it will not be necessary to change them if a vendor change or a router upgrade is necessary. The only change will be to the router configuration templates.

So, how can we create a network definition that will last longer and can be used as the source of truth for the entire network automation? Let's explore a few points that would help with that.

Nested and hierarchical definitions

A network definition does not need to be a flat unique file definition but can use a group of files in a nested setup. The reason is that some definition files can be specific to a particular characteristic present in all devices, such as vendor, device type, device rules, ACLs, or device function. Subsequent files that follow the hierarchy can have details that are more specific, such as location, name, capacity limits, or size.

Using nested or hierarchical network definitions will help to avoid having large definition files for each device, and most important of all, avoid the repetition of definitions across different files.

For instance, imagine that you want to control a list of IP addresses that are allowed to log into all routers in your network. If you don't use nested definitions, you might need to add the list of IPs to all router definitions. But if you have nested definitions, you might be able to use only one file definition for that.

To use hierarchy and groups, though, you must create a custom library that compiles the final definition of one specific router by looking into all hierarchical definition files that belong to that router. The final compiled definition can then be used on the router configuration render to complete the router template and output the correct router configuration.

IP allocation considerations

One important point to be observed with definitions is the IP addresses that are associated to every interface or protocol on the network devices. The IP addresses are normally unique to the network, unless **Network Address Translator** (**NAT**) is being used, but the majority of the IP address range is unique per region and per device.

To create a more flexible and future-proof solution, the IP allocation has to be as unfixed as possible, and rules can be taken for the configuration render to allow better use of the IP allocation.

Using an IP allocation engine in combination with nested definitions, it is possible to reserve IPs that are essential for device identification, such as loopbacks, and leave other IP ranges to be associated to network interfaces.

One enhancement that can be made when IP addresses are not fixed is to have a service that translates the IP address to a name associated with the network definition files. That can be done using DNS, for instance.

The strategy of having less-fixed IPs in the definitions will allow a more flexible solution and avoid complications for the network definition files.

Using files for definitions

When creating definitions, the best practice is to use plain files, not databases or any other storage method. This will allow engineers to have a complete source of truth without dependency on any system or application, thus the files can be read even after a catastrophic event with multiple system failures.

File format

Network automation should use only one file format across all network definitions. The file should be text-based but use a well-known format that enforces, among other things, typing. If the structure of the file is standard and easy to read, it will help engineers to review it if necessary.

Names

Whenever associating names to network definitions, avoid acronyms or any sort of abbreviation. Although shortening can help engineers to type faster, it can create confusion and raise problems when humans have to investigate. Remember that you can always create hotkeys or aliases locally in your environment to type device names faster.

Some devices have a limit on string sizes for names, so use names wisely and describe as much as possible the device you are naming to avoid economizing space.

This section has shown how important it is to examine some details when creating network definitions. We'll now review the most common data representation types in our network definitions.

Exploring different file types

The examples that we have seen in this chapter had network definitions with files ending with `.yaml`, which is an indication that they are in YAML format. But why? Why not XML or JSON formats? Let's explore the pros and cons of the most used file formats.

For our network definitions, we want to choose a format that is easy to read by humans and systems, fast to parse, and small to store. Then we can write a large content of definitions in files without worrying about performance, reading difficulty, or storage issues.

XML files

Extensible Markup Language or **XML** is the oldest markup language of those described in this section, with the first implementation dated 1996. Its first standard publication was created in 1998 by the World Wide Web Consortium as the version 1.0 specification.

The main goals of the initial XML design were to create a markup language that was simple, could cover general cases, and was easy to use across the internet. Even though the initial idea was for XML to create documents, the language has been used for arbitrary data structures on client and server interactions.

Because of its schema system, XML can use several media types. In 2001, IETF published the RFC3023, which described all possible media types, including `application/xml` and `text/xml`. In 2014, IETF published RFC7303, which refined the standards for media types and rendered RFC3023 obsolete.

Here is an example:

```
<router>
  <interface>
    <ip>
       <unicast>
       10.2.2.3
       </unicast>
    </ip>
  </interface>
  <loopback>
```

```
    <ip>
        <unicast>
        100.1.1.1
        </unicast>
    </ip>
    </loopback>
</router>
```

Note that in the example, we only have two data values (the highlighted IP addresses); the rest is markup overhead.

Here are the pros and cons:

- Pros:

  - More flexible for representing general data

- Cons:

  - Slow to process because of its complexity

  - Lots of overhead from using many repetitive structure markers

  - Less human-readable

  - Is prone to redundancies in structure

JSON files

JavaScript Object Notation, or **JSON**, is a newer method for data interchange representation compared to XML, with the original specification having been done by Douglas Crockford in the early 2000s. In 2004, IETF published an informational RFC4627, but only in 2014 did IETF create the RFC7159 standard. Now the latest standard is RFC8259.

JSON, like objects in JavaScript, has primitive types such as string, Boolean, number, and null. The structure in JSON consists of a key name and a value surrounded by a pair of curly brackets, as in {"key": <value>}. The key name is always a string. Here is an example:

```
{
    "router": {
        "interface": {
            "ip": {
                "unicast": "10.2.2.3"
```

```
        },
    "loopback": {
      "ip": {
        "unicast": "100.1.1.1"
      }
    }
  }
}
```

Note that JSON carries much less overhead than XML.

The preceding representation can also be written in one line, but it is not easy to read:

```
{"router": {"interface": {"ip": {"unicast": "10.2.2.3"}},
"loopback": {"ip": {"unicast": "100.1.1.1"}}}
```

Here are the pros and cons:

- Pros:

  - Simpler and faster than XML

  - Better parsing performance

  - Loading truncated files is avoided

- Cons:

  - Does not support comments

  - Does not allow aliases

  - Carries more overhead than other formats

  - Is not human-readable, depending on the format

YAML files

YAML was originally an acronym for **Yet Another Markup Language** because it was created after the proliferation of markup languages such as XML and HTML in late 1990. The creator Clark Evans wanted YAML to sound different, so the name was changed to **YAML Ain't Markup Language**, a recursive acronym, to differentiate the purpose of YAML from other markup languages. The standard version 1.0 was published in 2004 and the latest version, 1.2.2, was published in 2021.

YAML is intended to be human-readable and its data representation requires the use of indentation and new lines, which are used for delimitation and the grouping of data; this is different from JSON, which doesn't actually require newlines or indentation.

YAML also supports advanced features that are not supported by other data representation languages, such as anchors and references, which are very useful for avoiding repetition and data errors. Natively, YAML encodes scalars (such as strings, integers, and floats), dictionaries (or maps), and lists.

Here is an example:

```
router:
  interface:
    ip:
      unicast: 10.2.2.3
  loopback:
    ip:
      unicast: 100.1.1.1
```

As you can see, YAML has an even shorter representation and is very easy to read. As opposed to JSON, though, it can't be represented in just one line as the format alters the data representation.

Here are the pros and cons:

- Pros:

  - Simpler and smaller

  - Easy to read

  - Allows aliases and anchors

  - Allows comments

- Cons:

  - Not as fast to parse as JSON

There are other formats, such as **TOML**, **HOCON**, and **HCL**. Each of them has its advantages and disadvantages, but for our network automation and for most of our network definitions, YAML is the best option so far. It is also the most common one in network definitions.

Summary

In this chapter, we explored how network automation and engineers can benefit from having a proper configuration and network definition solution. It is not easy to grow a network with minimal human interaction and low dependency on network vendors and operating system versions.

You are now familiar with network configuration issues and how to tackle them. You are able to distinguish the stages of a deployment pipeline. You are also able to create a robust network definition to feed an automated router configuration render and choose the best file type to represent the network definition.

The next chapter will cover network programming by looking at what we should and should not do when writing code for networks.

Part 2:
Network Programming
for Automation

The second part of the book is focused more on the programming aspects of network automation. This includes a description of the popular libraries, runtime performance, scaling aspects, error handling, logging, and more. Go and Python are used, and, in some cases, there are comparable examples to show how either language can be used for better network automation work.

This part has the following chapters:

- *Chapter 5, Dos and Don'ts for Network Programming*
- *Chapter 6, Using Go and Python for Network Programming*
- *Chapter 7, Error Handling and Logging*
- *Chapter 8, Scaling Your Code*

5

Dos and Don'ts for Network Programming

Writing code for networks is exciting because when it fails, it is challenging, and when it works, it is rewarding. If you are an experienced programmer, you will have an easier ride, but if you are a newbie, it will be stormy. Let's dive into some coding practices that will help you get through these storms easier.

We are going to focus on this chapter on coding aspects for Python and Go related to network programming. The subject covered here would be also good for any type of programming; however, we are going to focus on programming for networks, and these are the topics we will cover:

- Coding topics
- Applying best practices in coding
- Coding formatters
- Versioning and concurrent development
- Testing your code

At the end of this chapter, you should be familiar with coding terms used by the community and which of them matter most. You will be able to understand coding best practices and how to become a better network code developer. If you are an experienced coder, it will be a good refresh. If you are new, this chapter will be your mantra when writing code.

Coding topics

Writing code used to be very simple and straightforward; it only required the ability to understand a program's workflow, its performance, and its algorithms. But today, the story is a bit different. Coding now has a culture that has evolved in the last few decades. What matters most is the code's reusability and, therefore, its style. To be reusable, code has to be easy to understand and should have few or no bugs or security issues.

If you are new to coding, or network coding, it is important to know all the topics used in coding culture today. Let's discuss briefly the most important ones in this section.

Peer review

It's not recommended to write code and publish without a **peer review**. Peer review allows the coder to be consistent with the team and avoid undesirable mistakes. However, for most organizations, this process can be slow and sometimes expensive. One alternative is to use software robots that can perform most of the peer review that was once done by another software engineer.

A peer review is a conscious and objective review done by another developer that has the objective of checking team standards, language standards, and community best practices, sharing knowledge, verifying code repetition and name convention, and aligning design with implementation, among others. As the list is long, the peer review process is not an easy and straightforward job; it sometimes takes longer to review the code than to write it.

Life cycle

It is important to keep track of your code by measuring and storing the date of creation, compilation (if any), and distribution. This is done by having some sort of life-cycle management. With a life cycle, it is possible to trigger actions to go back to the source code and verify whether the current recommendations are still valid on the source code, and also check whether there is any new security vulnerability. There are tools that do life-cycle management automatically to your code base and your applications; use them when possible.

Refactoring

In computer programming, **refactoring** (or **code refactoring**) is the term used to change the code without changing its external behaviors in terms of output and input. Refactoring can be used to fix bugs, improve performance, remove security vulnerabilities, or comply with new code styling. The important thing here is that refactoring does not change the features of the code. In other words, the external behavior, inputs, and outputs are preserved.

> **Important note**
> Check more on code refactoring on the site `https://refactoring.guru/`.

Copying code and licensing

The first thing you should do before publishing your code is to think about the **copyright** and **code license**. Why? Because if your code is good, it will definitely be reusable by someone else or will be copied for other purposes, unless you strictly specify the rules of copying on the license document. In addition, if you are importing or using external libraries, you might be compromised by copyright issues.

Therefore, it is important to create appropriate license documents and read the license documents of the code you are using or copying. The types of licenses that are attributed to software are listed here:

- **Trade secret licenses**: No information made public; private internal usage; unpublished
- **Proprietary licenses**: Have copyrights; no public licenses and no source code
- **Noncommercial licenses**: Used for noncommercial use and no source code
- **Copyleft licenses**: Grant use rights; forbids relicensing and public source code
- **Permissive licenses**: Grant use rights; allows relicensing and public source code
- **Public domain licenses**: All possible grants

Why do we care about these types of licenses? Because you are not supposed to distribute or publish the software you are working on. You could receive legal penalties and fines for breaching the license agreement on the code you are creating locally at your company. Most of the time, you are probably going to reuse someone else's library or code into a major piece of software. Before using an external library or code, the best way is to consult the legal department and check whether there are any licenses that are forbidden for use internally.

Consequently, it is recommended when writing code to check the license disclaimer document provided. It is normally located in the root directory of the source code, in capital letters, with names such as LICENSE or LICENSE.rst, but sometimes COPYRIGHT, COPYING, or something similar is used to express the copy grants.

For open source, the code also can rely on rules defined by definitions created by other entities; to do that, you just need to read the LICENSE file and check which one is applied. One of the most used ones in open source code is the **MIT License**.

Before using the open source library or copying it into your code, make sure the license does not present a risk to your organization. The open source code that presents the most risk when reusing it in private organizations is **GPL 2.0** and **GPL 3.0** (**GNU General Public License**). The low-risk ones are **MIT License**, **Apache License 2.0**, and **BSD License 2.0**.

The following table shows the top five most used licenses in projects at GitHub:

Rank	License	% of projects
1	MIT	44.69
2	Other	15.68
3	GLP 2.0	12.96
4	Apache	11.19
5	GLP 3.0	8.88

Table 5.1 – Rank of the most used licenses in GitHub projects

The following screenshot shows the evolution of GitHub project licenses from 2008 to 2015. Note the growth of MIT licenses:

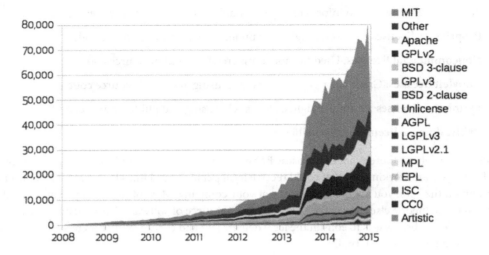

Figure 5.1 – License usage in GitHub projects (source: github.blog)

Now that you are familiar with the most popular code licenses, let's progress with discussing code quality and perception.

Code quality and perception

Code quality is a term normally heard among software engineers, but to be very clear, it is not an easy subject, and the definition of quality is normally subjective and prone to personal preferences.

To avoid such an ambiguous definition, code quality should focus on team or company guidelines and standards. If the company that you have just started has loose or no code guidelines, the code *quality* probably relies on the senior engineer. But that must—and should—be avoided.

Code quality should be measured by using standards and code practices, not by a person. If there are no standards or code practices in your team, use this chapter to guide you while creating one.

When defining the quality of the code, try to use measurable metrics that will allow software engineers to understand poor quality without taking it personally. Look for the following characteristics in the code:

- **Reliability**: Measure how many times the code runs without failures
- **Maintainability**: Measure how easily the code can be changed, including size, complexity, and structure

- **Testability**: Measure how the code can be controlled, observed, and isolated to create automated testing such as unit tests (which will be discussed in the *Testing your code* section later in this chapter)

- **Portability**: How easily the code can run in different environments

- **Reusability**: Measure how the code can be reused by dependencies or by copying

When possible, use automated tools that can classify and format the code, which avoids personal subjective analysis.

Architecture and modeling

Software architecture and design is a very long and wide topic. Depending on the size of your project, you might not need to do a design or architecture when working with network automation code. However, it is important to know they exist and can be used effectively to have a better code structure and better organization.

I would suggest that the most important part of software architecture is the modeling, which helps a lot in the early stages to validate the structure and the organization of the code. Communication with a model is easy with customers and can be understood even if you are not a software developer.

The most popular used models are **Unified Modeling Language (UML)** as generic, **Systems Modeling Language (SysML)**, a subset of UML, **Service-oriented architecture Modeling language (SoaML)**, also a subset of UML, and the **C4 model**. It does not matter which one you will choose; using any model before you start coding will help you to get the code right.

Let's use the deployment pipeline described in *Chapter 4* as an example to create a software model of the solution. *Figure 5.2* illustrates how it would be a very simple model for the deployment pipeline:

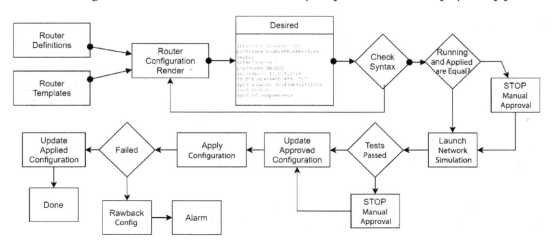

Figure 5.2 – Deployment pipeline software model example described in Chapter 4

The preceding diagram illustrates, using arrows and boxes, how a developer would tackle the problem of constructing a deployment pipeline. The inputs are the router definition and templates, then a configuration render creates the configuration and stores it at the desired state. The syntax is checked, and then the running config is verified if it is equal to `applied`. If not, it stops and waits for manual intervention; if it is okay, then launch a network simulation. Then, all tests are done. If the tests are passed, update the approved configs. The next step is to apply the configuration to the routers, and if it is applied correctly, then update the `applied` configs.

We have discussed in this section a few code culture topics that are used nowadays within the developing community. In the next section, we are going to dive into what the best practices are for writing code for Python and Go.

Applying best practices in coding

Here are points that are not necessarily standards, but they are important when we are writing code for networks.

Follow the standards

In the '90s, the internet was starting to grow fast, and programmers started to become more present in the community. Shell scripting started to grow as a viable tool to help system engineers work with their servers. Then came **Perl** as a powerful scripting language to help as well. Other languages were also mature, such as C, C++, and Java. But every developer wrote differently with their own style, which did not help code collaboration and made sharing code a disaster. When Python was created, the coding style problem already existed. Something had to be done to avoid confusion when sharing code.

Let's check now what the standards are for Python and Go languages.

Python standard style

Python can be written in so many different ways. If you are an old programmer, you would remember Perl, which was a scripting language like Python but without any particular style, which was a mess. There are documents that help programmers to write better code in Python. The documents are created by the Python community, and they are called **PEPs**, which stands for **Python Enhancement Proposals**.

Initially, the PEP process was created to propose and discuss consensus on new features, but it is also used to document information. A PEP is similar to an internet **Request For Comments** (**RFC**) for Python, and when describing new features, it has to contain a description of the feature in detail and a rationale for the creation of it.

When proposing new features for Python, the writer is responsible for building consensus within the community, describing different options, collecting input from the community, and documenting it.

Some PEPs are informational and are basically used to guide Python coding style. If you are planning to become a Python code developer, a PEP is like your daily journal to check and read. It is important to keep updated with the latest PEP and make sure your code follows the recommendations.

So far, there are a few useful PEPs to help the community with guidance and best practices on how to write code with Python. Let's check the most important ones.

Writing code with the guidance of PEP-8

PEP-8 is perhaps the most important PEP you have to read if you are new to Python because it describes in detail all styles and conventions when writing code in Python. Although the language does not force you to select a style when writing, using *PEP-8* (or *PEP8*) will make your code easy to understand for anyone in the Python community.

When you are new to Python, it might be difficult to remember what a part of the code you wrote weeks or a few days ago was supposed to do. However, if you follow *PEP-8*, you can be sure that you have named your variables well, you have added enough whitespace, and you have commented your code well. Following these guidelines will make your code more readable for others and for you when coming back to read after a while. Beginners will also learn Python much faster and with ease when following *PEP-8*.

The *PEP-8* document is very clear; it shows a wrong code style and then a correct style with an explanation of why to use that style.

Go standard style

Before describing the standards in the Go language, it is important to note that Go is also known as **Golang** because its former website was hosted on `golang.org`. Now, the Go language is hosted on `go.dev`.

Go is quite different from Python in terms of code formatting and style. Go does not have PEP-like documents to point out how to write better code in Go. So, the style standard relies mostly on the internal formatting tool, called `gofmt`, which is responsible for formatting the source code in a predictable way.

Although the `gofmt` formatter can be skipped, it is highly recommended to use it as soon as you change something in your code and before publishing it. Some tools will run the formatter automatically before running or compiling your Go program. In addition to `gofmt`, there are Go **linters** that can point out style errors or suspicious constructions; we are going to describe these in the *Coding formatters* section later in this chapter.

Furthermore, the Go community has published several guidelines that help with writing better code. Here is a list of the best ones:

- `https://github.com/uber-go/guide/blob/master/style.md`
- `https://github.com/dgryski/awesome-go-style`
- `https://staticcheck.io/docs/checks/`

Mindful code writing

Most of the time, we are going to be reading code instead of writing because the process of writing code drives you to read other parts of the code. So, you know where your code goes, how your code follows the existing design, and the behaviors the existing code has so that it doesn't crash. The important point here is, even if you have no intention to share your code with anyone else, it will be probably forgotten by you in a few months, and if you might need to update it, you will have a hard time understanding what it does. So, write always thinking about someone reading it, even if this is just you.

Therefore, the process of writing code must be much more mindful than the process of reading. When someone reads part of the code, it should not create any doubts; it should be straightforward and easy, and quick to comprehend.

Making it extremely readable

When writing code, don't save words or phrases. If you have the opportunity to choose a mistake when writing your code, choose the one that over-explains your code. Why? Because your code should be understandable by anyone, and that includes you, in a few months or years. Therefore, make sure your code is easy to understand and easy to follow by someone new.

Commenting your code

The code should be clear to understand; however, sometimes, it is necessary to comment on your code to help readers and reviewers understand your code. You should add comments throughout the code when you want to emphasize nuances in your code, describe details of some algorithms, or warn the reader about something odd.

Comment headers

Some text headers in comments are used to help the reader avoid spending time on code that should or might be changed. When you start coding with your team, check which terms are used for helping code readers and code reviewers. The comment headings listed here are not fixed and change between organizations and teams. Here are the most popular comment headers:

- TODO: Perhaps the most common comment used is TODO, which is used to designate part of the code that needs to be added or refactored. This comment helps the reviewer avoid spending time reviewing a chunk of code that needs to be modified.

- FIXME: Some developers when writing code also use the FIXME comment heading, which designates the part of the code that is wrong, ugly, has performance issues, or is too complicated. Use FIXME when you know something is wrong and should be fixed soon.

- HACK: This text heading is used to designate part of the code that has a workaround to fix a bug or to help performance. The HACK heading is also another important comment to show the reader a quick workaround to be fixed later, helping them understand why that part of the code is there.

- BUG: This heading is used to show a problem in the code, which must be fixed soon. The author has identified the problem and decided to comment on that part of the code as a bug. The code reviewer can then evaluate and decide whether they have to accept the code with the bug or ask for corrections. Again, these comments save the time of the reviewer as they do not need to evaluate that part of the code in depth. Use this whenever you see a problem in the code.

- NOTE: This heading is for the author to communicate with the readers some notable gotchas or particular details to help readers understand the code faster. Use them wisely to help reviewers understand, as too many NOTE headings can also be annoying and create a distraction.

Docstring

A docstring refers to a comment string that is used in Python to describe a module, function, class, or method. Docstrings are particular to Python and are detailed in *PEP-257*.

Some **integrated development environments** (**IDEs**) such as PyCharm and IntelliJ will automatically create a docstring for you as soon as you define a function.

Here's an example of a multiline docstring for a Python function:

```
def router_connection(hostname, timeout):
    """Perform a connection to a router
    :param string hostname:
    :param int timeout:
    :return: True | False
    """

    print("Connecting to host", hostname)
    .
    .
    .
    return True
```

Note that the docstring is always initiated and ended with triple-double quotes (" " "). The preceding example is a suggestion and other styles are possible. To visualize all possible styles, read the docstring conventions in *PEP-257*.

> **Note**
> Docstring documentation can be found at `https://peps.python.org/pep-0257/`.

Godoc

In Go, there is `godoc`, which is simpler than a docstring as it is not a language construct or machine-readable syntax; it is just string text comments. The convention is simple: to document a variable, function, package, or constant, just write comments directly before its declaration. Here is an example of the comments used for the `Asin` function in the Go `math` package:

```
// Asin returns the arcsine, in radians, of x.//
// Special cases are:
//
//     Asin(±0) = ±0
//     Asin(x) = NaN if x < -1 or x > 1
func Asin(x float64) float64 {
    if haveArchAsin {
        return archAsin(x)
    }
    return asin(x)
}
```

Note that in this example, the comments are placed before the `Asin` function, and they describe in detail what to expect with the function and its special cases. Comments are always text that starts with a double slash (//).

The preceding example was taken from the Go source code at `https://github.com/golang/go/blob/master/src/math/asin.go#L14-L25`.

After commenting your code in Go, add explanations on functions such as in the preceding example. Developers can use the `godoc` tool, which extracts text and generates documentation in HTML or TXT format.

> **Note**
>
> The `godoc` documentation is available at `https://pkg.go.dev/golang.org/x/tools/cmd/godoc`.

Use IP libraries

Newbies in programming for networks may think IP addresses can be treated as a string or a list of characters. But when dealing with networks, it is very important to use the correspondent IP library to load an IP from a text file or to create one from a loop. There are several reasons you might want to avoid strings for IPs; some are presented here:

- Avoid numbers that are out of the scope of an IP, such as the number 256
- Avoid malformed text IP addresses
- Make sure not to overlap other space addresses
- Easy to find the network, mask, and broadcast
- Easy transition from IP version 4 to IP version 6

Let's check how to use these IP libraries in Python and Go.

IP libraries in Python

The IP address library in Python did not exist natively until version 3.3. Before version 3.3, Python coders had to use external libraries to handle IP addresses. Python had to incorporate this library internally after realizing its importance. The transition was documented in *PEP-3144* (`https://peps.python.org/pep-3144/`).

Examples of IP manipulation using Python

First, let's print all valid IP addresses in a subnet:

```
import ipaddress

network = ipaddress.ip_network("10.8.8.0/30")
print(list(network.hosts()))
```

Here's the output:

```
[IPv4Address('10.8.8.1'), IPv4Address('10.8.8.2')]
```

Note that the output type is `IPv4Address` and not a string. The `ipaddress.ip_network()` method automatically detects whether it is an IPv4 or IPv6 address.

Now, let's print the subnets that exist in a larger subnet, excluding one subnet:

```python
import ipaddress

network = ipaddress.ip_network("10.8.8.0/24")
subnet_to_exclude = ipaddress.ip_network("10.8.8.64/26")
print(list(network.address_exclude(subnet_to_exclude)))
```

Here's the output:

```
[IPv4Network('10.8.8.128/25'), IPv4Network('10.8.8.0/26')]
```

Note that the type on the output is not an IP address type as it was before, but a network type called IPv4Network.

IP libraries in Go

Go is different from Python as IP address libraries came after the Go language's first version. IP manipulations in Go are faster than in Python because Go is not interpreted like Python but compiled.

Examples of IP manipulation using Go

In Go, we are first going to check whether an IP address belongs to a subnet:

```go
import (
    "fmt"
    "net"
)

func main() {
    _, subnet, _ := net.ParseCIDR("10.8.8.0/29")
    ip := net.ParseIP("10.8.8.22")

    fmt.Println(subnet.Contains(ip))
}
```

Here's the output: `false`.

The 10.8.8.22 IP address does not belong to the 10.8.8.0/29 subnet as the program prints false.

The next example checks whether a subnet belongs to another subnet:

```go
import (
    "fmt"
    "net/netip"
)

func main() {
    first_subnet := netip.MustParsePrefix("10.8.8.8/30")
    second_subnet := netip.MustParsePrefix("10.8.8.0/24")

    fmt.Println(second_subnet.Overlaps(first_subnet))
}
```

Here's the output: `true`.

In this example, `first_subnet` belongs to `second_subnet`, so they overlap. Note that in this example, we used a `"net/netip"` import instead of `"net"`, which is a newer IP manipulation package introduced to the Go language.

In terms of comparison, the Python built-in library has more features for IP manipulation than the built-in Go IP packages. On the other hand, Go can manipulate IPs faster.

If you are looking for particular features for IP manipulation that are not present in `net` or `net/netip`, you can use a newly developed community package called `inet.af/netaddr`, described at `https://pkg.go.dev/inet.af/netaddr`.

Follow naming conventions

It is wise to verify with your team what the local development name conventions are before starting to publish any code for review. There are particular differences that might be a problem to use in your organization. Here, we are going to describe the most popular name conventions used, but feel free to adapt this to your local team culture.

Before starting to describe the name conventions, let's define three ways of writing multi-word names, as listed here:

- **Camel case**: Each word, except the first, starts with a capital letter—for example, `myFirstTestName`

- **Pascal case**: Each word starts with a capital letter—for example, `MyFirstTestName`

- **Snake case**: Each word is separated by an underscore character—for example, `my_first_test_name`

In the case of defining constants, all characters in the name must be in uppercase for most programming languages. Therefore, snake case is used always in lowercase, unless the name is describing a constant. Here's an example of a name describing a constant that will define the maximum storage capacity: `MAX_STORAGE_CAPACITY`.

Naming in Python

For Python, there are a few rules to follow when writing variable names. Here is a short list of what to do:

- Variable names can start only with letters or underscore characters (_)
- The variable name can only have alphanumeric characters and underscores
- Never start a variable name with a number
- Dashes (-) are used only for package and module names, never for variables
- Double underscore characters starting in variables are reserved for Python
- Use pascal case for class names
- Use snake case for module names and functions
- Constants in Python use all uppercase characters—for example, `MYCONSTANT`

> **Note**
>
> Google has a created a great name convention guide for Python, available publicly at `https://google.github.io/styleguide/pyguide.html#316-naming`.

Naming in Go

For Go, there are a few rules to follow. Here is a list of the major ones to follow:

- Variable names use camel case or pascal case
- As with variable names, function names use camel case or pascal case
- Variable or function names don't start with a number
- Use only alphanumeric characters for variable and function names
- Dashes or underscores are normally not used for names
- Go allows the use of snake case for variable names, but check whether your organization's name convention allows it, because it is not common to use snake case in the Go community
- Constants in Go are like in Python: use all uppercase characters

Moreover, Go has a particular difference from other languages, and uses uppercase in the variables inside a Go package (or module). So, if the variable name starts with uppercase, it is normally a variable that can be accessed from outside the Go package. When the variable starts with lowercase, it's a local variable that is only accessible with code within the same Go package.

> **Note**
>
> A good reference for names in Go can be obtained at `https://go.dev/doc/effective_go`.

Don't shorten variable names

In the early times of computer programming, variable names had to be small to save more space in memory when saving the uncompiled code. But today, this is not a problem, so don't save memory when writing variable names.

If you have to write a variable for representing the number of birds flying, write `number_of_birds_flying` in snake case or `numberOfBirdsFlying` in camel case.

There is an exception to this rule if you are using variables inside list comprehension or variables in a loop such as counters or indexes; then, shortened variables are okay to use.

Avoid complex loops

Loops are added to code for several reasons: to increment, interact with a list, repeat an operation, and read a stream, among others. It is possible to use loops for deterministic or non-deterministic sizes, which can be very useful when you don't know what you are going to interact with.

But irrespective of the intention you are using a loop in your code for, use them wisely so that they don't become unreadable or too complex to understand. A good rule of thumb is to avoid too many lines inside a loop. I would say that 20 lines would be the limit. Another good practice is to avoid nested loops or loops within loops. If you have to do a nested loop, limit the number of levels to no more than two. Loops are hard to read when they are nested, and even worse when there are too many lines inside them.

So, what is the best way to create a loop with fewer than 20 lines? The suggestion is to group a few operations into functions that can be called inside the loop. Name the function whatever you want it to do, and add the operations to it. When you are reading the loop, the function call will be easy to spot and the name of the function will indicate what the operation is doing, which is really easy to read and understand. Also, it is easier to write unit tests when using this approach.

How to avoid nested loops? Use functions to call a loop. Instead, you can also use generators, list comprehensions, or maps to avoid loops. Using maps has the advantage of making your code even faster as you don't need to interact with it. List comprehensions are used in Python. Generators can be used in Python and are implemented using goroutines in Golang. So, whenever you need a nested loop, try not to overcomplicate it, and, if possible, use one of these mechanisms to avoid it.

Don't repeat code

In software development, there is one term that is called **DRY** or **Don't Repeat Yourself**. The worst a developer can do is repeat code. Yes, it is bad—really bad. Why? Because if you write code that performs a task over and over in multiple places, whenever you want to modify that task, you will be required to carry out the same change in several places. The other problem is that you will have to write multiple unit tests for the same task.

The opposite view of DRY is **WET**, which stands for **Write Everything Twice**. A WET solution might occur because of multi-tiered architecture where developers have different tasks, and in the end, repetition may happen. A DRY approach should eliminate code redundancy by using a framework that helps developers avoid repetition.

In this section, we have seen some examples of how we can use best practices when writing code. In the next section, we are going to explore how we can use tools to help us write better code.

> **Note**
>
> There are many more suggestions on best practices when writing code. A more extensive list can be found at `https://github.com/golang/go/wiki/CodeReviewComments` and `https://go.dev/doc/effective_go`.
>
> For Python, I recommend the book *Effective Python* by Scott Meyers.

Coding formatters

As we can now see, there are several rules and conventions to follow when writing code in Go or Python, and the list is long, which does not help when you are a beginner. But thanks to the community, there are tools that help to fix the code format for you. These tools can automatically fix your code or suggest modifications. Let's have a look at a few code formatters for Python and Go.

Python Black

Black is a very popular code formatter for Python. It is *PEP-8* compliant, so it checks all *PEP-8* recommendations. The default run will reformat the code to be *PEP-8* compliant, but you can run it as a preview to identify the code change suggestions.

With Black, it is possible to fix Python code automatically. Here is an example of wrong code written for Python:

```
if long_variable_name is not None and \
  long_variable_name.field > 0 or long_variable_name.is_debug:
  z = 'hello '+'world'
else:
  world = 'world'
  a = 'hello {}'.format(world)
  f = rf'hello {world}'
```

After running Black, the code will look like this:

```
if (
    long_variable_name is not None
    and long_variable_name.field > 0
    or long_variable_name.is_debug
):
    z = "hello " + "world"
else:
    world = "world"
    a = "hello {}".format(world)
    f = rf"hello {world}"
```

Black does apply all recommendations of *PEP-8* and also has its own style. In the preceding example, Black has correctly changed the `if` statement and added the correct indentation. In addition, it changed the strings assignment from single quotes to double quotes. There is no specific PEP designating if you have to write strings with double or single quotes, but Black's own style uses double quotes for strings.

I personally think code that has single quotes in some strings and double quotes in other strings looks very ugly. So, I would recommend using double quotes only for strings, as Black does.

> **Note**
> More on Python Black can be found here: `https://github.com/psf/black`.

Python isort

One important bit that is not actually explained easily are the import lines that are present in the Python code. Sometimes there are dozens of imports, and if they are not properly organized, it is difficult to realize which ones are used and which are not.

Thanks to the `isort` utility, it is possible to fix `import` statements, group them correctly, and sort them automatically.

Here is one example of code that has a loose ugly `import` statement:

```
from central_lib import Path
import os
from central_lib import Path3
from central_lib import Path2
import sys
from my_lib import lib15, lib1, lib2, lib3, lib4, lib5, lib6,
lib7, lib8, lib9, lib10, lib11, lib12, lib13, lib14
import sys
```

After running `isort`, the code looks like this:

```
import os
import sys

from my_lib import (lib1, lib2, lib3, lib4, lib5, lib6,
                     lib7, lib8, lib9, lib10, lib11, lib12,
                     lib13, lib14, lib15)

from central_lib import Path, Path2, Path3
```

Note that `isort` grouped the imports and sorted them in a predictable way. If you need to add extra imports, you will not repeat or miss them.

> **Note**
> More on `isort` can be found here: `https://pypi.org/project/isort/`.

Python YAPF

Although *PEP-8* guidelines may allow you to write code in a way that other developers will appreciate, it does not necessarily mean your code looks good. The **YAPF** acronym is not actually explained on the tool's page, but it perhaps means **Yet Another Python Formatter**.

In essence, YAPF is similar to Black, which has its own style besides the *PEP-8* recommendations, but a major difference is that it can be configured for fine-tuning style formatting.

YAPF is based on `clang-format` developed by Daniel Jasper, which takes the original code and reformats it using the YAPF style guide, even if the original code did not violate any *PEP-8* guidelines. If used in all code, the style remains consistent throughout the project, and reviewers would not need to discuss or argue about the style during code review.

> **Note**
> More on YAPF can be found here: `https://github.com/google/yapf`.

YAPF, `isort`, and Black are the three major Python code formatters, but the list of Python formatters is very extensive. Other formatters can be found at `https://github.com/life4/awesome-python-code-formatters`.

Go gofmt

The Go language has its own formatter that comes with the Go language package, and it is called `gofmt`. Its objective is similar to Python Black and Python YAPF, which formats the source code to the best format that can be done consistently. The formatter is used with the CLI command of `go`—for example, `go fmt myprogram.go`.

Some platforms run automatically `gofmt` every time you want to build or run a Go program.

Go golines

`golines` is a formatter that shortens long lines in addition to fixes done by `gofmt`. This formatter was created because `gofmt` does not break long lines and it is hard to visualize when lines are too long.

Here is an example before running `golines`:

```
myVariable := map[string]string{"a key": "a value", "b key":
"b value", "c key": "c value", "d key": "d value", "e key": "e
value"}
```

After running `golines`, the code looks like this:

```
myVariable := map[string]string{
        "a key": "a value",
        "b key": "b value",
        "c key": "c value",
        "d key": "d value",
        "e key": "e value"
}
```

More details on how `golines` works are available at `https://yolken.net/blog/cleaner-go-code-golines`.

Go golangci-lint

A programming language lint or linter is software used to point errors, bugs, style errors, and suspicious constructs in the source code. It does not fix these automatically but flags errors and warnings to be fixed. For Go, the best linter package is `golangci-lint`, described in detail at `https://golangci-lint.run/`. The original linter used for Go was located at `https://github.com/golang/lint` but was deprecated and frozen because of a lack of contribution.

There are several checks that `golangci-lint` can do. The default most important checks are listed here:

- `govet`: Checks whether format strings in `Printf` calls have correct aligned arguments
- `unused`: Checks unused variables, functions, and types
- `gosimple`: Points to parts of the code that can be simplified
- `structcheck`: Verifies unused struct fields
- `deadcode`: Verifies any code that is not being used

A complete list of linters available can be found at `https://golangci-lint.run/usage/linters/`.

> **Note**
> A list of additional Go code formatters can be obtained at `https://github.com/life4/awesome-go-code-formatters`.

There are many more awesome tools that can format your code, not only via command-line tools but also via integrated IDE formatting tools. This section was created to introduce some of the tools used by Python and Go and show how to use them to help your code improve. In the next section, we are going to talk about how to use tools when developing code to help concurrent development and versioning.

Versioning and concurrent development

Nowadays, writing code is not only about the lines of code you are creating but also about the code others are writing to contribute to your software. In addition, code that has changes requires some information regarding why the changes are there and some version tags to identify these so that developers can easily revert changes or use different versions when testing or deploying.

How can we accomplish such tasks when writing code? The best answer is to use a version control system. Today the most popular free tools are **Git**, **SVN**, **Mercurial**, and **CVS**. Git, for sure, is the most popular of all, basically because of Linux and the growth of the GitHub website. *Table 5.2* shows a quick comparison between these four version control systems.

Version system	Year created	Famous for	Website
Git	2005	Linux and `github.com`	`www.git-scm.com`
SVN	2000	FreeBSD and `sourceforge.com`	`subversion.apache.org`
CVS	1986	NetBSD and OpenBSD	`www.nongnu.org/cvs/`
Mercurial	2005	Python and Mozilla	`www.mercurial-scm.org`

Table 5.2 – Most popular version control systems comparison

When writing code, it is recommended to use a source version control system, which allows the software to be edited by multiple developers at the same mechanism to avoid adding code to the system without syncing the changes. Several different terms are used in versioning systems; the terms are almost the same over different systems such as Git or SVN.

Let's describe the most common commands used by versioning systems.

clone

`clone` is a command to copy the whole source code, including subdirectories and versioning system data, to your local directory.

Here's an example of the command being used in Git: `git clone https://github.com/brnuts/matrix.git`.

checkout

`checkout` is a command to update files in the working directory to match the source tree source code. It is also used to switch to another code branch.

Here's an example of the command being used in Git for updating the current branch: `git checkout`.

And here's an example of it being used in Git for checking out another branch: `git checkout mybranch`.

commit

While changing the source code is recommended to change small chunks, each small change should attach a text tag describing the changes. The `commit` command is a command that adds a text tag to the code and a checkpoint number. As your code changes, you add commits to each small change. After the whole change is complete, you will have a change log that has multiple commit text explanations, helping the reader understand why and how the changes were made.

Mainline

Mainline is normally the name of the main **branch** of the code. Other branches of the code normally ramify from the main branch or from other branches, which after ramification have a parent branch. In this sense, the mainline is a branch that has no parent branch.

The mainline branch is known as the trunk of the source code tree and can also be called **master** or sometimes **baseline**.

Branching

You can use versioning systems to create branches in your code. A **branch** is normally a ramification of the code from another branch. The goal of having a branch is to make particular changes to the code without interfering with the parent branch.

The code branch can then have as many changes as needed, and when ready can be added to the parent branch by using the `merge` command.

There are several advantages of using branches—one of them is to clarify the communication on the intention of a large change by adding several small change **commits** on the branch, with an explanation of each change. Another advantage is to allow other developers to continue to add changes to the parent branch.

A branch without a parent branch is the mainline branch or the master branch.

Merging

A code **merge** is used when a branch is ready to be added to the main branch. The `merge` command checks whether code can be added without interfering with the mainline branch code. If there is interference, it will create code **merge conflicts** that need to be resolved before the branch can be merged successfully. The `merge` command can also be done between individual branches without a mainline.

The merging technique is the safest technique used to combine branches without compromising any code that differs from other code. But sometimes, there are so many conflicts to be resolved that this might not be feasible. When that occurs, it is recommended to start a new branch and try to add the changes, gradually merging eventually, or to use the `rebase` command, which allows the commit to be applied from one branch to another branch interactively if necessary.

Have a look at the following diagram:

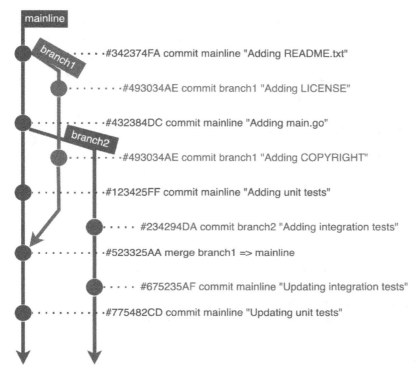

Figure 5.3 – A description of the mainline commits and branches

Figure 5.3 shows an example with all details of the versioning system for the mainline and the branches. You can also spot the commits and merges that are done for each branch and for the mainline as well. Note that branch 1 (green) was created to add two files, called LICENSE and COPYRIGHT, then was merged to the mainline. Branch 2 was created from a 432384DC commit and has not been merged to the mainline yet.

In this section, we investigated how versioning tools help a great deal when developing code concurrently. Use a versioning system whenever you can to organize your development. In the next section, we are going to explore why and how we add testing to our code.

Testing your code

Testing code is another subject that has grown in the past decades. In the early '70s, testing code was not actually part of the development process but part of the debugging process when there was a problem with the software. Testing assumed significance when American computer scientist Glenford Myers published the classic book *The Art of Software Testing* in 1979, proposing the separation of debugging from testing.

Since then, much has improved in the development process with testing. Today, it is about preventing errors, and the process to develop code with tests that prevent errors is called **test-driven development** or **TDD**.

In the process of developing the code, the developer must identify two parts of the code: **testable** and **non-testable** code. Testable code is pieces of code that are easily verified by low-level testing, such as a unit test. Non-testable code is pieces of code (such as the import of a library) that are not easily tested at a lower level and normally are tested with higher-level testing such as integration tests.

Therefore, when writing code, the developer must be mindful of when and how tests would be performed for that part of the code. The developer will easily identify which part of the code can be done with lower- or higher-level testing. There are several levels of testing. Let's discuss a few of them in the following subsections.

Unit testing

Unit tests are the lower level of testing because they are intended to cover a small part of code, which is the smallest piece of code that can be logically isolated. These pieces of code can be subroutines, functions, methods, loops, or properties. The goal of unit tests is to show that individual parts of the code are correct and—when used in methods or functions—to make sure they work as expected.

If in the future, code has its libraries or modules upgraded or refactored, the unit tests created before are able to identify whether the code still works with the change, making it robust against eventual bugs introduced with upgrades. For that, unit tests have to be written to cover all functions, modules, methods, and properties; then, whatever change happens that could cause a fault would be quickly identified by the unit tests.

Unit tests can be written in several ways: some would write a function and then write the unit tests for this function; some would write the unit tests and then write the function. When unit tests are written first, this is normally called **test-driven development** (**TDD**) or **extreme programming** (**XP**).

As some code cannot be tested with unit tests, each team and organization has a **test coverage** parameter used to evaluate each software package. The number varies, but normally it is higher than 50% test coverage for the software package, and something higher than 80% for all new functions and methods introduced.

Unit tests are easy to create when a function or method has clear input parameters and some outputs, but it is not easy to create when a function interacts with external elements such as a database. For dealing with these external environments, some languages have **mock capabilities** (or **mockups**) to simulate the external behavior, without doing it.

If part of the code is not unit-testable (mocks are not available), it will probably be covered by higher-level testing, as with the ones we are going to talk about next.

Integration testing

Integration tests are used to fill the gaps of tests not possible to be performed with unit tests. These are normally tests that touch or require external environments such as databases, external files, or service APIs.

From a testing order perspective, integration testing is the second step after the unit tests have been done successfully and before the **end-to-end** (E2E) testing (if required). Integration testing normally uses broader code coverage in terms of modules and functions. It groups them into a large aggregate and applies a test plan to validate the expected outputs. Individual functions and modules are tested using unit tests. During integration tests, they are not necessarily tested individually but in a higher-level group.

It is important to set up a test environment that can support integration testing, such as an external test database. During integration tests, the mockups added in the unit tests are not used, but real external testing environments are used. The external testing environment is normally separated from production and is dedicated to integration testing.

When performing integration tests, the test environment should behave in the same manner as the production environment, or at least very close to production. With similar production behavior, we can have more confidence, after the integration tests are passed, that the software will work in production.

Unit tests can run on someone's computer; on the other hand, integration tests can't because they rely on a testing environment that would validate the external communication from the software. Someone might say, *set up an environment for integration testing on your computer*, but it would then not be possible to validate some tests that require consistency.

E2E testing

Most tests can be done by unit tests and integration tests; however, some might still be missing. **E2E tests** are added in another level of testing, which will evaluate the overall software, including performance, under a production-like environment.

If possible and allowed, the ideal way is to run the E2E testing in the production environment. If it is not possible to run this in production, an isolated production-like environment can be used. The goal is to simulate the real scenario of the software running from the start to the end so that passing tests can validate the software and can be used without restrictions in production.

In terms of testing time and cycle, the E2E test should run last, as these tests take much longer to run than the integration tests. The test coverage should cover broader behavior, not only touching external interfaces but also using the whole workflow process that the software belongs to.

To summarize unit tests, integration tests, and E2E tests, let's use an example. Imagine your software creates a robot that can paint cars. What would the responsibility of each testing level be? Here's a possible implementation:

- **Unit tests**: Lower-level testing of small modules of the software:

  - Move x, y, and z directions
  - Self-identify unalignments
  - Identify which color a container has
  - Spray paint evenly in all directions
  - Don't leak paint
  - Identify the ending paint in the container

- **Integration tests**: Higher-level testing with external modules:

  - Can paint a car full
  - Can refill paint in an empty paint container
  - Can paint car with different colors
  - Doesn't leave blank color spots in the car

- **E2E tests**: These would be the higher level, ultimate tests:

  - Car can be painted in X minutes
  - No interference on the next car or previous car when painting the current car
  - Robot can paint Y number of cars per hour, with paint refill and color change

As we can see in this example, the number of tests is normally higher when the level of testing is lower. So, unit tests would have more tests than integration or E2E tests. However, the time to run tests is slower as we go higher up the levels. E2E testing should take longer to finish than integration or unit tests.

Other testing

There are several other testing classifications. For our network automation, we are going to use only the three levels we just described: unit tests, integration tests, and E2E tests. The others we are going to explain here, for reference, can be added to your automation work if you think they are necessary.

Let's talk briefly about them now.

System testing

Technically speaking, all tests can be done by unit tests, integration tests, and E2E testing. The **system testing** is to add an extra testing phase that can be used to evaluate the system's compliance with the specified requirements. Therefore, system testing is another name for testing in a formal procedure to verify that all requirements have been met.

Acceptance testing

When dealing with projects that involve end customers, it is important to align the expectations of the behavior of the software. Acceptance testing is a mechanism to negotiate and communicate with the end customer about the correct behavior of the system. **Acceptance tests** are like a contract between the end customer and the developer to delineate the final delivery of the software. The end customer is normally involved when selecting a test plan and tests so that the tests cover the system's behavior and expectations.

The usual terms used here are **user acceptance test** (or **UAT**) and **operational acceptance test** (or **OAT**).

If your network automation code is highly dependable on a customer's use, with clear requirements, I would highly recommend writing test cases with the customer for UAT and OAT. But if your work is for internal usage and has no end-user requirements, integration and unit tests are sufficient.

Security testing

If your software is deployed in restricted or vulnerable areas or touches network devices in a sensitive area, you might need to run security testing. Different from the other tests, **security testing** focuses on vulnerabilities of the code that you have created and also on all module dependencies that are imported by your code. The versions of all modules are also checked, and some versions might be prohibited because of some internal security policies.

One additional test is called a **pentest** or a **penetration test**, which is performed on the system your software is created on. The idea of a pentest is to perform a simulated cyberattack on the final system used by your code. The test has the objective of identifying weaknesses and potential access for unauthorized people to data or the system.

Security testing also verifies the password management in the code and any cryptography used. There are some best practices when using passwords, such as cryptography and secrets that reduce the risk of unauthorized people accessing the system.

For our network automation, security testing would be required if the devices we are going to automate were located in a vulnerable area, or if the software might be exposed to external parties or to the internet. But in general, there is no need for security testing for internal usage.

For sure, there is more to discuss in terms of security testing, but that would probably require a book on its own.

Destructive testing

Destructive software testing or **DST** is performed in the software to evaluate the robustness of the final system with the code that has been added. The tests are intended to stress the system or to enter invalid inputs so that the software starts to fail. The success of the test is to expose possible design weaknesses and performance limitations under normal and strained conditions.

Failures during destructive tests are seen as good results if the system shows clearly the methods it has used to recover or to avoid breaking down completely. The performance is measured under the test, and the system resources are stressed to the maximum until degradation or failure occurs. It is possible to use the measurements taken during the performance stress later to add input system limitations for the current implementation, consequently avoiding any degradation.

Alpha and beta testing

When tests are performed before going to production, some would call them **alpha** or **beta testing**. Normally, alpha tests are performed before beta tests, just because of the alphabetical order. The tests are not necessarily different from integration or E2E testing, but normally they are performed in a production-like environment, or sometimes in production.

Some beta tests also involve customers that are aware of the new features being tested and can help evaluate the new software before it goes to other customers in production. Therefore, beta testing would be the final stage before the software is considered reliable enough to be deployed in full production. Beta and alpha tests can have windows of testing that can last from hours to days.

Summary

Congratulations! At this point, you are familiar with the coding practices used today. You know how to use tools to help your source code in terms of quality and concurrent development. You are also able to perceive the importance of having unit tests and integration tests added to your code.

From now on, as a new code developer for automation, you are more equipped with the jargon and terms used in the software development community. You are able to improve not only the quality of your code but also your team's code quality. Use the tools and the topics discussed here throughout your automation development career.

In the next chapter, we are going to have more practical examples of how to use Go and Python for network automation. We are going to explore some examples of automation in Go and Python and compare them.

6

Using Go and Python for Network Programming

In this chapter, we're going to see how Python and Go are powerful and used for network programming, but depending on what your requirements are and your environment, one might be better suited for you than the other. We are going to use Python and Go for network programming by checking the advantages and disadvantages of using each.

By the end of this chapter, you will be able to identify which language (Python or Go) is more suitable for your network project and which library to use. You are going to learn the differences and superpowers of each language that will probably make the difference in your network automation work.

The topics we are going to cover in this chapter are as follows:

- Looking into the language runtime
- Using third-party libraries
- Accessing network devices using libraries

Technical requirements

The source code described in this chapter is stored in this book's GitHub repository at https://github.com/PacktPublishing/Network-Programming-and-Automation-Essentials/tree/main/Chapter06.

The examples in this chapter were created and tested using a simple network device simulator. The instructions on how to download and run this simulator are included in the Chapter06/Device-Simulator directory.

Looking into the language runtime

After writing your code and saving it, you are going to run it somewhere in your network. Go and Python have different ways to combine your source code and all imported libraries before running. Which one suits you more? Are there any relevant differences that are important to know? We'll discuss that in this section.

What are compiled and interpreted languages?

After writing your code, some computer languages need to be compiled to run on your machine, though some don't as they are interpreted line by line as it runs.

The languages that are compiled have to have a compiler that translates the source code into a series of bits and bytes that can run on the CPU architecture of your computer; it also has to link all static and dynamic system libraries. For instance, a computer with an Apple M1 processor will have a different compiler than an Apple with an Intel x86 processor. The result after the compilation is a binary program that can't be read by humans and when it runs, it is loaded from disk to main memory.

Once you have compiled, you don't need to have your source code to run your program. The machine that runs your code does not need the compiler or the source code, only the compiled program binaries, which adds free space, code privacy, and code security.

On the other hand, interpreted languages (as it stands) are interpreted by a code interpreter, which interprets your code line by line when it is running. These interpreted languages are also known as **scripting languages**. The machine that runs the interpreted language needs to have the interpreter and the source code, which unfortunately exposes the source code and needs additional space to store.

Examples of compiled languages include Go, C, C++, Haskel, and Rust. Examples of interpreted languages include Python, PHP, Unix Shell, JavaScript, and Ruby.

Java is a special case because it has a compiler but compiles to its own **Java Virtual Machine** (**JVM**) architecture, which is not the CPU architecture where the program will run. Once compiled, you can use it anywhere but will need to install a JVM for the specific CPU architecture, adding extra storage and runtime complexity.

Python interpreter

The Python interpreter is sometimes called a Python virtual machine as a reference to a JVM, which can run anywhere. But Python does not provide a virtual machine like Java – it provides an interpreter, which is quite different.

In Java, this virtual machine is like a virtual CPU that provides an environment for the Java bytecode-compiled program to run. The JVM translates the Java-compiled bytecode into the bytecode of the CPU architecture where it is running. So, Java code will need to be compiled first; then, the compiled program can run on any machine that has a JVM installed.

On the contrary, the interpreter is much more complicated as it does not translate bytecode as it does in a JVM but interprets lines in a context with its surroundings. The interpreter reads the whole code and parses the syntax that must be decoded in the program's context. Because of this complexity, this can be a very slow process.

Let's investigate some of the available Python interpreters.

Using CPython

CPython is one of the most common interpreter programs that is required to be installed in the machine where the Python code will run. CPython is written in C and is perhaps the first implementation of the Python interpreter.

CPython is the place where new functionalities will be created before they can be exposed in Python. As an example, when concurrency was added to Python, it was first achieved by the CPython interpreter process using the operating system's multitasking properties.

CPython implementations can be compiled into proprietary bytecode before being passed to the interpreter. The reason is that it is easier to create an interpreter based on a stack machine instruction set, even though an interpreter doesn't need to do so.

The following is an example of a CPython stack machine instruction set:

```
$ pythonPython 3.9.4   [MSC v.1928 64 bit (AMD64)] on win32
>>> import dis
>>>
>>> def return_the_bigest(a, b):
...     if a > b:
...         return a
...     if b > a:
...         return b
...     return None
...
>>> dis.dis(return_the_bigest)
  2           0 LOAD_FAST                0 (a)
              2 LOAD_FAST                1 (b)
              4 COMPARE_OP               4 (>)
              6 POP_JUMP_IF_FALSE       12

  3           8 LOAD_FAST                0 (a)
             10 RETURN_VALUE
```

```
4      >>     12 LOAD_FAST            1 (b)
              14 LOAD_FAST            0 (a)
              16 COMPARE_OP           4 (>)
              18 POP_JUMP_IF_FALSE   24

5             20 LOAD_FAST            1 (b)
              22 RETURN_VALUE

6      >>     24 LOAD_GLOBAL          0 (Null)
              26 RETURN_VALUE
```

As you can see, the `return_the_bigest` function is translated into the CPython bytecode shown, which will be used by the CPython interpreter when it's run. Note that the instruction set does the same as the `return_the_bigest` function, which is harder for humans to read and easier for the Python interpreter.

More on disassembler Python bytecode can be found here: `https://docs.python.org/3/library/dis.html`.

More on CPython can be found here: `https://github.com/python/cpython`.

Using Jython

Jython is another Python interpreter that was created originally in 1997 by Jim Hugunin as **JPython**. In 1999, JPython was renamed to Jython, as it is known today.

Jython is used to compile Python code into a Java bytecode virtual machine that can run on any hardware that has a JVM installed. Sometimes, it can run faster as it does not need to be interpreted like CPython.

Although the project started with high expectations, today, it only supports Python 2.7, and the support for Python 3.x is still under development. Therefore, you are only going to need Jython if you are running your code on a machine that only supports JVMs. There will also be lots of limitations as it only supports Python version 2.7, which is no longer supported by the Python community and was deprecated in January 2020.

More on Jython can be found here: `https://github.com/jython/jython`.

Using PyPy

PyPy is another Python interpreter implementation that claims to run Python code faster than CPython. PyPy also claims to handle concurrency better than CPython with the usage of micro-threads. And finally, it claims to use less memory than CPython.

Despite the great advantages of PyPy, CPython is still the most used Python interpreter, mainly because people don't know about PyPy and the default Python installation uses CPython.

PyPy has a website dedicated to comparing its speed with other interpreters, such as CPython. The website also has comparisons with other versions of PyPy. *Figure 6.1* shows a comparison between CPython and PyPy with information taken from the PyPy speed website. On average, PyPy is 4 times faster than CPython:

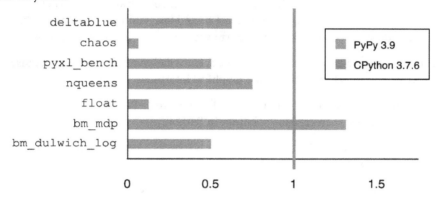

Figure 6.1 – CPython and PyPy comparison taken from speed.pypy.org

The preceding figure contains blue bars, which represent how a specific benchmark runs in PyPy 3.9 in comparison to CPython 3.7.6. As an example, the bm_dulwich_log benchmark runs twice as fast compared to CPython (0.5). The sixth blue bar shows that PyPy runs slower for the bm_mdp benchmark (1.3), but for the second blue bar, which represents the chaos benchmark, PyPy runs up to 20 times faster. Details about each benchmark can be obtained at https://speed.pypy.org/.

For more on PyPy, check its home page at https://www.pypy.org/.

Using Cython

Although some people compare PyPy with **Cython**, Cython is not a Python interpreter like PyPy. Cython is a compiler that can be used with the Python and Cython languages, which is based on **Pyrex**, a superset of the Python language. Cython can add C and C++ extensions easily to your code. As it is implemented in C, the Cython code claims to be faster than Python when using PyPy.

Therefore, if you are looking to write Python code and require high performance, try Cython. More on Cython can be found at https://cython.org/.

Using IPython and Jupyter

The major advantage of an interpreted language is that it can run interactively easier. Python has an interpreter for this called **IPython**.

IPython can be used to run your code gradually line by line and then check what happens in memory. This is useful when testing or trying a new code or function. It is also quite handy to get results as the program runs and adapts the code to suit your desired output during code development.

In conjunction with IPython, you can use a **Jupyter** notebook, which is a web interface that's easy to use and has graphical output capabilities.

As an example, imagine you need to gather information from 100 network nodes on CPU usage on the network and make a graphical report for the last hour. How could you do that quickly without worrying about building or testing? The best platform you can use is Jupyter notebook with IPython.

IPython and Jupyter are also frequently used for data science and machine learning, because of their advantages in terms of interactive methods and graphical interfaces.

For our network programming, IPython is a powerful tool to create **Proof of Concept** (**PoC**) coding and test new functionalities when creating solutions in Python.

More on IPython can be found at `https://ipython.org/`. More on Jupyter can be found at `https://jupyter.org/`.

With that, we have looked at the major Python interpreters. Now, let's look at how Go works.

Go compiler

In Go language development, there is no code interpretation like in Python code; instead, there's compilation. This compilation is done by the Go compiler, which normally comes with the Go language package. The compiler reads the source code and then translates it into the bytecode of the CPU architecture where this program is going to be executed. When executing the compiled code, there is no need to have the compiler or the source code, only the compiled binary code.

Because of this compilation, Go programs run faster than Python-based programs – in some cases, they can run 30 to 100 times faster, especially when dealing with concurrency. As an example, the `fannkuch-redux` benchmark, which is used to test multicore parallelism, takes 8 seconds to run in Go, whereas it takes 5 minutes to run in Python (source: `https://benchmarksgame-team.pages.debian.net/benchmarksgame/performance/fannkuchredux.html`).

Although the Go language distribution provides the compiler, the Go community has started other projects as alternatives for the Go compiler. One of them is called **TinyGo**, which is used when the compiled code does not have much memory to be stored, such as in small microcontrollers or small computers. Therefore, TinyGo is used when memory space in the running target computer is limited. More on TinyGo can be found at `https://github.com/tinygo-org/tinygo`.

Now, let's compare both languages in terms of computing runtime.

Pros and cons of programming runtimes

Let's explore the pros and cons of using Go and Python for programming while focusing on the code that will run on the machine during runtime.

Pros of using the Python runtime:

- Easy to create PoC code by using interactive Python with IPython and Jupyter notebooks
- Easy to create data visualization during prototyping
- Has a large community with different interpreters, libraries, and frameworks

Cons of using the Python runtime:

- Consumes more space and memory on the target running machine
- Consumes more CPU and is slower to complete tasks compared to Go
- Code is visible on the target running machine, which can be unsecure
- Runtime parallelism implementation is weak compared to Go

Pros of using the Go runtime:

- Consumes less memory and space on the target running machine
- Consumes less CPU and runs faster than Python
- Code is compiled, which is not human readable and can't easily be decoded
- The runtime parallelism implementation is much better than Python

Cons of using the Go runtime:

- More difficult to create prototypes
- A smaller development community and fewer runtime libraries

For network programming, Go has much more benefits compared to Python in terms of performance. However, as Python is an older language, it has a broader community with more network libraries and functionalities.

So, choosing which language to use will depend on the use case you are working on. If you want something quick and wish to write less code by reusing someone else's library, then Python is probably the best fit. But if you want something that has more performance, is secure, uses less memory, and can be built as one binary program, Go is your best fit. As the Go community grows, it will probably have more libraries that can help with network automation over time, but as it stands today, Python has more community contributions in the network automation field.

In the next section, we'll learn how we can add external libraries to our code in Go and Python.

Using third-party libraries

When developing network automation, it is always worth researching tools and libraries in the community to see if you can incorporate some external code that will add functionalities or speed up your development process.

In preparation to explain how to use third-party libraries, it is important to understand how libraries in general are used in Python and Go. We'll detail the library-adding process in Python and Go in this section.

Adding Python libraries

Before discussing how libraries are added to Python, it is important to explain that, in Python, a **library** is also known as a **package** or a **module**. These three terms are used widely throughout Python documentation, which can make some new Python developers confused. So, whenever you see the term "Python library", it can also mean a package or a module.

A library in Python can be external, internal, or built-in. These external libraries are also known as third-party modules, packages, or libraries.

To use a library in Python, you just need to use the import statement at the beginning of the code. If the library is not found, it will raise an error exception called ModuleNotFoundError, as shown here:

```
$ python
Python 3.10.4 (main, Apr  8 2022, 17:35:13) on linux
Type "help", "copyright", "credits" or "license" for more
information.
>>> import netmiko
Traceback (most recent call last):
  File "<stdin>", line 1, in <module>
ModuleNotFoundError: No module named 'netmiko'
```

In the preceding example, the Python interpreter threw an exception called ModuleNotFoundError, which means the package is not installed or is not in the search path. The search path is normally included in the path variable inside the sys package, as shown in the following example:

```
$ python3.10
Python 3.10.4 (main, Apr  8 2022, 17:35:13) on linux
Type "help", "copyright", "credits" or "license" for more
information.
>>> import sys
>>> sys.path
```

```
['', '/usr/lib/python3.10', '/usr/lib/python3.10/lib-dynload',
'/usr/local/lib/python3.10/dist-packages', '/usr/lib/python3/
dist-packages']
```

Note that in the preceding example, the `sys.path` variable has been pre-filed with a list of system paths, but you can append more if needed.

Now, let's discuss how the built-in, standard, and external modules are used in Python.

Using Python built-in libraries

The built-in modules (or libraries) are modules that can be imported, but they come within the Python runtime program. They are not external files to be added, so they don't need to be found in the `sys.path` variable. There are also built-in functions, such as `print`, but built-in modules are the ones that you are going to explicitly import before using them, such as the popular one known as `sys`, or others such as `array` and `time`. These built-in modules are not external programs like they are in Python standard libraries, but they are included in the binary code of the interpreter, like in CPython.

Using Python standard libraries

These are modules that come with the Python distribution, but they are separate Python files that are added when you state `import` at the beginning of the Python code; they need to be found in the `sys.path` variable. These modules are Python programs and can be found in the Python library installation directory, such as the `socket` library, as illustrated in the following example:

```
$ python3.10
Python 3.10.4 (main, Apr  8 2022, 17:35:13) on linux
Type "help", "copyright", "credits" or "license" for more
information.
>>> import inspect
>>> import socket
>>> inspect.getfile(socket)
'/usr/lib/python3.10/socket.py'
```

Note that the `/usr/lib/pytho3.10/socket.py` location will depend on the operating system and version of Python. In the preceding example, Linux Ubuntu and Python 3.10 are being used. A list of all standard libraries for Python 3.10.x can be found at `https://github.com/python/cpython/tree/3.10/Lib`. More information about each built-in and standard library for Python 3.x can be found at `https://docs.python.org/3/library/`.

Using third-party Python libraries

External libraries or third-party libraries are modules (or packages) in Python that are not included with the Python distribution and need to be installed manually before they're used. These modules are normally not maintained by the Python library team but by developers of the Python community, which are not necessarily related to the Python main distribution.

As we discussed in *Chapter 5*, before adding the external module to your code, have a look in the `LICENSE` file and check if any limitations might impact the usage of this module on your code or in your organization.

The Python community has organized a group called **Python Packaging Authority** (**PyPA**), which is responsible for documenting and creating the infrastructure for Python developers to create and publish Python external packages. The group is responsible for maintaining a set of tools that are used to create, install, and maintain Python packages. The most popular tool is called `pip`.

PyPA is also responsible for maintaining `pypi.org`, where all packages that can be included with `pip` are documented. On this site, there is a search engine for finding packages and also documentation for developers that want to contribute or share their packages. Note that the source code for the packages is not on the `pypi.org` site, but in repositories such as **GitHub**.

Now, let's go through an example of the process of using a third-party package in our code. The package that we are adding as an example is called `netmiko`:

1. Check if the package is included on `pypi.org` as per PyPA. If so, we can use `pip` to add the package to our local Python environment.

    Yes, it is: `https://pypi.org/project/netmiko/`.

2. Read the `LICENSE` file and check if it is allowed to be used in your organization.

    This license is based on MIT, which is less restrictive, so we can use it: `https://github.com/ktbyers/netmiko/blob/develop/LICENSE`.

3. Use `pip` to install the package to your local Python environment, as shown in the following example:

    ```
    $ pip install netmiko
    Installing collected packages: netmiko
    Successfully installed netmiko-4.1.0
    ```

4. Check if you can import and locate where the `netmiko` library is installed:

```
>>> import netmiko
>>> import inspect
>>> inspect.getfile(netmiko)
'/home/claus/.local/lib/python3.10/site-packages/
netmiko/__init__.py'
```

In the preceding example, the `netmiko` package has been installed using the `pip` tool, and the library is located under my home directory, `/home/claus`. However, this varies, depending on the version of Python and the operating system used. It also varies within Linux, depending on the distribution, such as Debian, Ubuntu, or Red Hat.

Just keep in mind that third-party Python libraries will be installed normally under a directory called `site-packages`. Here are some examples of where it can be located for each operating system:

- For macOS: `/Library/Frameworks/Python.framework/Versions/3.9/lib/python3.9/site-packages`

- For Windows: `C:\Users\username\AppData\Local\Programs\Python\Python39\lib\site-packages`

- For Linux Ubuntu: `/usr/lib/python3.9/site-packages`

A complete list of all packages that you have installed on your system can be found by typing the following command:

```
pip list -v
```

Now, let's explore how we can use libraries and third-party libraries in Go.

Adding Go libraries

In contrast with Python, Go only has two types of libraries: the standard libraries and the third-party libraries. For Go, there is no concept of built-in libraries because it does not have an interpreter; instead, it has a compiler. The reason is that an interpreter can include a few libraries in its binary, which are called built-in libraries.

To add a library to your Go code, you will need to use the `import` statement, which is the same in Python but the syntax is a bit different.

Similar to Python, in Go, a library is also known as a **package**.

Let's have a look at how to add standard libraries and then third-party libraries in Go.

Using standard libraries in Go

Each version of Go includes a set of standard libraries that are installed with the Go distribution. These standard libraries are also Go programs that, when imported, will be combined with your code during compilation.

A list of standard libraries can be found at https://pkg.go.dev/std. The website is very useful because it includes an explanation of each library (or package).

The standard libraries can also be found in your local development filesystem. Go's installation includes all standard libraries, such as the fmt and math packages. The location of these standard libraries will vary for each operating system but can be found by looking into the GOROOT environment variable.

Most operating systems do not populate the GOROOT variable, so they will use the Go language's default location. To find out your default location, you can run the go env command, as shown here:

```
$ go env GOROOT
/usr/lib/go-1.19
```

In the preceding example, GOROOT is located in /usr/lib/go-1.19.

To learn how the libraries are located, let's use the fmt standard library to print a string to the computer terminal:

```
package main
import "fmt"
func main() {
    fmt.Println("My code is awesome")
}
```

In the preceding example, the import statement tells the Go compiler that it needs to add the fmt package, which in this case is a standard library package. In this example, the Go compiler will search for this package, starting by looking in the /usr/lib/go-1.19 directory.

To be more specific, the fmt package is located at /usr/lib/go-1.19/src/fmt. The Println function that's being used in this example is described in the /usr/lib/go-1.19/src/fmt/print.go file.

In Go, all programs have to belong to a package, and the package name is described in the first line of the code. In the fmt directory (/usr/lib/go-1.19/src/fmt), all the files included in the directory have a first line that contains the package fmt statement. This includes the scan.go, format.go, and print.go files, which are located in the fmt directory.

Other examples of standard libraries in Go can be found in your local Go installation, normally under the src directory. In the preceding example, it is under /usr/lib/go-1.19/src. Some other examples include the math package, located at /usr/lib/go-1.19/src/math/, and the time package, located at /usr/lib/go-1.19/src/time/.

Here is another example, which uses the math standard library:

```
package main
import (
    "fmt"
    "math"
)
func main() {
    fmt.Println("Square root of 2 is", math.Sqrt(2))
}
```

In the preceding example, the math library uses the Sqrt function, which is described in the /usr/lib/go-1.19/src/math/sqrt.go file. Again, all the files in the math directory have the package math statement on their first line.

Now, let's learn how to add third-party libraries to Go.

Using third-party libraries in Go

Third-party library packages in Go are added in a similar way to standard libraries – by using the import statement in your code. However, the underlying process of adding these packages to the Go development environment is a bit different in terms of a few characteristics.

In the case of third-party packages, the compiler needs to search for new files containing the package in a different path, which is populated in the GOPATH environment variable. Like GOROOT, you don't need to populate GOPATH as the Go compiler has a default location for the GOPATH variable.

In our Go example, let's check the default location for GOPATH by running the following command:

```
$ go env GOPATH
/home/claus/go
```

As we can see, the default location for GOPATH is the go directory inside my home directory (/home/claus).

In Go, third-party libraries can be added by invoking the `get` subcommand on the `go` command line, as shown here:

```
$ go get golang.org/x/crypto
go: downloading golang.org/x/crypto v0.0.0-20220622213112-
05595931fe9d
go: downloading golang.org/x/sys v0.0.0-20210615035016-
665e8c7367d1
go: added golang.org/x/crypto v0.0.0-20220622213112-
05595931fe9d
```

In the preceding example, we added the `x/crypto` package. Note that all other dependencies of the package are also added; in this example, the other dependency is `x/sys`. The `get` subcommand also stores the version of the package – in this case, `v0.0.0-20220622213112-05595931fe9d`.

Then, the package is saved under the `GOPATH` directory. In the preceding example, it is saved at `/home/claus/go/pkg/mod/golang.org/x/crypto@<VERSION>`.

Every time you run `go get` and there is a new version available, Go stores the new version and keeps the old one in a different directory.

You don't need to run `go get` when adding a third-party library. The `get` command is sometimes invoked when you run `go build` if the package has not been downloaded or cached, as shown here:

```
$ go build
go: downloading golang.org/x/crypto v0.0.0-20220622213112-
05595931fe9d
```

In the preceding example, the Go program had an `import` statement:

```
import (
    "golang.org/x/crypto/ssh"
)
```

After adding a third-party library, Go also updates the `go.mod` file. This file is required to keep track of the versions of the packages that are added to your Go program. The following is the content of the `go.mod` file for the preceding examples:

```
$ cat go.mod
module connect

go 1.19
```

```
require (
    golang.org/x/crypto v0.0.0-20220622213112-05595931fe9d
)
require golang.org/x/sys v0.0.0-20210615035016-665e8c7367d1 //
indirect
```

Note that the go.mod file also stores the versions of the packages for all dependencies. In the preceding example, the x/sys package version is also stored in the file.

Now that you're familiar with adding third-party libraries to Go and Python, let's dive into the next section and look at some libraries that can be used to access the network.

Accessing network devices using libraries

So far, we have discussed how to run and work with libraries in Python and Go. Now, let's focus on how to use Python and Go to access network devices, which is one of the most important points in our network automation work.

In *Chapter 3*, we discussed several methods to access network devices. One of the most popular ones is using a **command-line interface** (**CLI**). We also discussed SNMP, NETCONF, gRPC, and gNMI. In this section, we are going to explore a few examples of how to use libraries to access network devices, mainly using the CLI. Later, we are going to explain and show the libraries for accessing the network using other methods.

Libraries to access the network via a CLI

There are lots of libraries on the internet that can access network devices, and some of them might be obsolete or not used anymore. Here, we are going to present the most popular ones in chronological order, from the old to the newer ones.

The following examples are going to collect the uptime of the network device by sending the uptime command via the CLI on an SSH connection.

Using Python Paramiko

Paramiko is perhaps one of the oldest implementations for accessing network devices via the CLI. The first release was published in 2003. Today, it has more than 300 contributors and there are almost 2,000 packages depending on Paramiko (https://github.com/paramiko/paramiko/network/dependents).

As we discussed in *Chapter 3*, the correct way to use CLI libraries is via **Secure Shell** (**SSH**). Paramiko implements secure cryptography for SSH using a lower-level library called PyCrypto (http://pycrypto.org/).

Let's look at a simple example to obtain the uptime of the network host:

```python
import paramiko
TARGET = {
    "hostname": "10.0.4.1",
    "username": "netlab",
    "password": "netlab",
}
ssh = paramiko.SSHClient()
ssh.set_missing_host_key_policy(paramiko.AutoAddPolicy())
ssh.connect(**TARGET)
stdin, stdout, stderr = ssh.exec_command("uptime")
stdin.close()
print(stdout.read().decode("ascii"))
```

If the connection is successful, the output of the preceding program will be the uptime of the network host:

```
13:22:57 up 59 min,   0 users,   load average: 0.02, 0.07, 0.08
```

Using Python Netmiko

Netmiko is also another popular library for accessing network devices via the CLI. The first release was published in 2014 and was built on top of Paramiko to simplify connections to network devices. In summary, Netmiko is simpler than Paramiko and more focused on network devices.

Let's learn what the same example for Paramiko will look like when using Netmiko:

```python
import netmiko
host = {
    "host": "10.0.4.1",
    "username": "netlab",
    "password": "netlab",
    "device_type": "linux_ssh",
}
with netmiko.ConnectHandler(**host) as netcon:
    output = netcon.send_command(command)
print(output)
```

As you can see, the implementation of the Netmiko code is just 2 lines, with much smaller and simpler code compared to Paramiko. The big advantage of the Netmiko library is that it handles the device's command prompt automatically and works with ease when changing the configuration in privilege mode since the prompt normally changes. In the preceding example, the device type was `linux_ssh` because our target host was a Linux device.

Netmiko can support dozens of network devices, including Cisco, Huawei, Juniper, and Alcatel. A complete list of devices can be found here: `https://github.com/ktbyers/netmiko/blob/develop/PLATFORMS.md`.

Using Python AsyncSSH

AsyncSSH is a modern implementation of Python **AsyncIO** for SSH. Python AsyncIO was introduced in Python version 3.4 and allows Python to work concurrently using async/await syntax and provides high performance for network access. To use AsyncSSH, you will need to have Python version 3.6 or higher.

In summary, AsyncSSH provides better performance for multiple hosts, but it is a lower-level implementation that requires more complexity in your code to handle network devices in comparison to Netmiko.

The same example of getting uptime for AsyncSSH can be written like so:

```python
import asyncio, asyncssh, sys
TARGET = {
    "host": "10.0.4.1",
    "username": "netlab",
    "password": "netlab",
    "known_hosts": None,
}
async def run_client() -> None:
    async with asyncssh.connect(**TARGET) as conn:
        result = await conn.run("uptime", check=True)
        print(result.stdout, end="")
try:
    asyncio.get_event_loop().run_until_complete( run_client() )
except (OSError, asyncssh.Error) as execrr:
        sys.exit("Connection failed:" + str(execrr))
```

Note that when using AsyncSSH, you will have to work with routines and events, as instructed by Python AsyncIO and the preceding example. If you want to dig into this, some great documentation can be found at `https://realpython.com/async-io-python/`.

Using Python Scrapli

In comparison to Netmiko, Scrapli is newer. The first release was published in 2019 and is also built on top of Paramiko but has capabilities to use AsyncSSH, which claims to improve performance when accessing multiple devices. The idea of the name Scrapli came from *scrape cli*, similar to someone scraping the screen. This is because Scrapli's main objective is to interpret text from the network terminal using the CLI.

Scrapli was built to allow its users to interpret prompts from multi-vendor network devices such as Netmiko but with a reduced set of platforms. The 2022.7.30 version supports Cisco, Juniper, and Arista. One advantage of using Scrapli is that it also supports NETCONF.

The following is the same example to get uptime but using Scrapli:

```python
from scrapli.driver import GenericDriver
TARGET = {
    "host": "10.0.4.1",
    "auth_username": "netlab",
    "auth_password": "netlab",
    "auth_strict_key": False,
}
with GenericDriver(**TARGET) as con:
    command_return = con.send_command("uptime")
print(command_return.result)
```

So far, we have learned about the most popular libraries for accessing network terminals using Python. Now, let's learn we can do this with the Go language.

Using Go ssh

In contrast to Python, Go does not have a large community for working with libraries that can handle network devices. Therefore, in some particular cases, you might have to write your own scrape mechanism to interpret the remote network terminal via the CLI. The problem with doing this yourself is it will be different for each network vendor and will take more coding time as you add different network devices.

Let's look at an example of getting uptime from the remote network host:

```go
package main

import (
    "bytes"
    "fmt"
    "log"

    "golang.org/x/crypto/ssh"
)

func main() {
    host := "10.0.4.1"
    config := &ssh.ClientConfig{
        User:            "netlab",
        HostKeyCallback: ssh.InsecureIgnoreHostKey(),
        Auth: []ssh.AuthMethod{
            ssh.Password("netlab"),
        },
    }

    conn, err := ssh.Dial("tcp", host+":22", config)
    if err != nil {
        log.Fatalf("Dial failed: %s", err)
    }

    session, err := conn.NewSession()
    if err != nil {
        log.Fatalf("NewSession failed: %s", err)
    }

    var buff bytes.Buffer
    session.Stdout = &buff
    if err := session.Run("uptime"); err != nil {
        log.Fatalf("Run failed: %s", err)
```

```
        }

        fmt.Println(buff.String())
}
```

As you can see, the preceding example contains much more code than some higher-level libraries.

Using Go vSSH

There is a library in Go called vSSH, which is built on top of golang.org/x/crypto by engineers at Yahoo. It creates an abstraction for accessing remote terminals on network devices, which avoids the code that we saw previously with the SSH example.

One of the main claims of vSSH is that it can handle access to multiple targets with high performance, which is obtained by using Go routines (a great beginner's guide for Go routines can be found at https://go.dev/tour/concurrency/1).

Although vSSH can handle multiple targets efficiently, let's start by writing an example that uses only one target:

```
package main

import (
    "context"
    "fmt"
    "log"
    "time"

    "github.com/yahoo/vssh"
)

func main() {
    vs := vssh.New().Start()
    config := vssh.GetConfigUserPass("netlab", "netlab")
    vs.AddClient(
        "10.0.4.1:22", config, vssh.SetMaxSessions(1),
    )
    vs.Wait()

    ctx, cancel := context.WithCancel(
```

```
        context.Background()
    )
    defer cancel()

    timeout, _ := time.ParseDuration("4s")
    rChannel := vs.Run(ctx, "uptime", timeout)

    for resp := range rChannel {
        if err := resp.Err(); err != nil {
            log.Println(err)
            continue
        }
        outTxt, _, _ := resp.GetText(vs)
        fmt.Println(outTxt)
    }
}
```

Go routines have been used in this example. Additional targets can be added using vs.AddClient() before calling vs.Wait(). In our example, only one target was added to get the uptime of the remote network host. The loop at the end is not necessary for only one target, but I left it there to demonstrate how to use it with multiple targets.

As we can have one host that might be faulty or slow when using multiple targets, a parse timeout is used, which in our example is 4 seconds. The preceding example uses a channel via the rChannel variable to obtain the results of the Go routines for each target.

Using Go Scrapligo

Scrapligo is a Go version of the successful Python library Scrapli. It supports the same network platforms and also supports NETCONF. The advantages of using Scrapligo over Scrapli are the ones we discussed earlier in this chapter related to how Go and Python runtime perform.

The preceding example in Scrapligo would look like this:

```
package main

import (
    "fmt"
    "log"
    "github.com/scrapli/scrapligo/driver/generic"
```

```
        "github.com/scrapli/scrapligo/driver/options"
)

func main() {
    target, err := generic.NewDriver(
        "10.0.4.1",
        options.WithAuthNoStrictKey(),
        options.WithAuthUsername("netlab"),
        options.WithAuthPassword("netlab"),
    )
    if err != nil {
        log.Fatalf("Failed to create target: %+v\n", err)
    }

    if err = target.Open(); err != nil {
        log.Fatalf("Failed to open: %+v\n", err)
    }

    output, err := target.Channel.SendInput("uptime")
    if err != nil {
        log.Fatalf("Failed to send command: %+v\n", err)
    }

    fmt.Println(string(output))

}
```

We have just discussed how to access network devices via the CLI using Python and Go libraries. Now, let's learn how to use libraries to access the network using other methods.

Libraries to access networks using SNMP

Excluding the CLI, the second most popular network device access method is SNMP. But as we discussed in *Chapter 3*, SNMP is only used to read information from network devices. The SNMP write method is not used for the reasons discussed in *Chapter 3*.

For the SNMP examples, we are going to pick up one library from Python and one from Go. These libraries are the most popular ones for using the SNMP method today.

In the previous subsection, we used CLI methods to collect the uptime of the network device. We are going to demonstrate now that we can also get the uptime of the remote network device by using the SNMP method. For that, we need to collect the SNMPv2-MIB::sysUpTime MIB variable.

Using Python PySNMP

PySNMP is the most popular library for Python for the SNMP method. It supports all versions of SNMP from version 1 to version 3. The following is an example of using the SNMP method to obtain the uptime of the network device:

```python
from pysnmp.hlapi import *

snmpIt = getCmd(SnmpEngine(),
    CommunityData("public"),
    UdpTransportTarget(("10.0.4.1", 161)),
    ContextData(),
    ObjectType(ObjectIdentity("SNMPv2-MIB", "sysUpTime", 0)))

errEngine, errAgent, errorIndex, vars = next(snmpIt)

if errEngine:
    print("Got engine error:", errEngine)
elif errAgent:
    print("Got agent error:", errAgent.prettyPrint())
else:
    for var in vars:
        print(' = '.join([x.prettyPrint() for x in var]))
```

The output will be SNMPv2-MIB::sysUpTime.0 = 72515.

More on PySNMP can be found at https://pysnmp.readthedocs.io/en/latest/.

Using gosnmp

For Go, the most popular library for the SNMP method is `gosnmp`. It also supports version 1 to version 3 of the SNMP protocol. Compared to PySNMP, `gosnmp` is newer but has more developers and more users, making it more reliable in terms of future development. The following is an example of collecting uptime from a network device using the SNMP method in Go. In this example, the OID number (`1.3.6.1.2.1.1.3.0`) represents the same as `SNMPv2-MIB::sysUpTime`:

```go
package main

import (
    "fmt"
    "log"
    snmp "github.com/gosnmp/gosnmp"
)

func main() {
    snmp.Default.Target = "10.0.4.1"
    if err := snmp.Default.Connect(); err != nil {
        log.Fatalf("Failed Connect: %v", err)
    }
    defer snmp.Default.Conn.Close()

    //SNMPv2-MIB::sysUpTime
    oid := []string{"1.3.6.1.2.1.1.3.0"}

    result, err := snmp.Default.Get(oid)
    if err != nil {
        log.Fatalf("Failed Get: %v", err)
    }
    for _, variable := range result.Variables {
        fmt.Printf("oid: %s ", variable.Name)
        fmt.Printf(": %d\n", snmp.ToBigInt(variable.Value))
    }
}
```

The output will be `oid: .1.3.6.1.2.1.1.3.0 : 438678`.

More on gosnmp can be found at `https://pkg.go.dev/github.com/gosnmp/gosnmp`.

Libraries to access networks using NETCONF or RESTCONF

When writing configuration to the network devices, the preferred method would be NETCONF or RESTCONF. However, some devices or some functions of the device might not have it implemented yet. In this case, the most appropriate method would be via a CLI as SNMP is not used to write data on the device.

The RESTCONF/NETCONF methods are newer methods to access network devices compared to a CLI or SNMP. Because of that, there are not many libraries available. Today, the best library to use NETCONF in Python would be Scrapli; for Go, this would be Scrapligo.

Examples of using Python Scrapli with NETCONF can be found at `https://scrapli.github.io/scrapli_netconf/`.

Examples of using Go Scrapligo with NETCONF can be found at `https://github.com/scrapli/scrapligo/tree/main/examples/netconf_driver/basics`.

You can also use a plain HTTP library to collect information using RESTCONF, as shown in the following Python example:

```python
import requests
from requests.auth import HTTPBasicAuth
import json

requests.packages.urllib3.disable_warnings()

headers = {"Accept": "application/yang-data+json"}

rest_call = "https://10.0.4.1:6060/data/interfaces/state"

result = requests.get(rest_call, auth=HTTPBasicAuth("netlab",
"netlab"), headers=headers, verify=False)
print(result.content)
```

Now, let's learn how to use Python and Go to access networks using gRPC and gNMI.

Libraries to access networks using gRPC and gNMI

Compared to the other methods, gRPC is quite new and network device vendors have added this capability in recent years. So, if you have old devices in your network, you might not be able to use gRPC or gNMI.

As we discussed in *Chapter 3*, gRPC is a more generic method, and gNMI is more specific for network interfaces. The main use of gNMI is network telemetry by invoking the underlying gRPC streaming subscription capability. Using gNMI allows your network code to scale easily and collect much more network management data compared to SNMP. The gNMI libraries are built on top of the gRPC protocol.

All major network device vendors have some sort of gRPC and/or gNMI implementation on their newer network operating systems. Among them are Cisco, Juniper, Arista, Nokia, Broadcom, and others.

Using gRPC in Python

Only newer versions of Python support gRPC, and the Python version has to be 3.7 or above. To use it, you need to install `grpcio` (`https://pypi.org/project/grpcio/`).

Examples of using gRPC in Python can be obtained here: `https://grpc.io/docs/languages/python/quickstart/`.

Using gRPC in Go

In Go, gRPC can run on any major version. It is well documented and several examples can be found at `https://github.com/grpc/grpc-go`.

Using gNMI in Python

Python support for gNMI is particularly rare. There are not many libraries available in the Python community. The following list describes the major ones:

- `Cisco-gnmi-python` was created in 2018 and was initially supported by Cisco Networks. The library was created by Cisco to foment the use of gNMI on Cisco devices, and perhaps not a good match for multi-vendor support. More details can be found at `https://github.com/cisco-ie/cisco-gnmi-python`.

- `gnmi-py` was created in 2019 and is sponsored by Arista Networks. This library does not support multi-vendor platforms and can be only used for Arista devices. More details can be found at `https://github.com/arista-northwest/gnmi-py`.

- `pyygnmi` was created in 2020. The library can be imported using `pip` and has been tested on Cisco, Arista, Juniper, and Nokia devices. This would be the preferred choice for multi-vendor platform support. More details can be found at `https://github.com/akarneliuk/pygnmi`.

Using gNMI in Go

For Go, gNMI is more mature and has better support compared to gNMI implementations for Python. There is only one library that can use gNMI in Go, which is called `openconfig-gnmi`.

The `openconfig-gnmi` library was created in 2016/2017 by Google and is now supported under the GitHub `openconfig` group. More on this library can be found at `https://pkg.go.dev/github.com/openconfig/gnmi`.

Besides the `openconfig-gnmi` library, there are other libraries related to gNMI in Go that you might find useful. Here is a list of the major ones:

- `google-gnxi` is a combination of tools that can be used with gNMI and gNOI. Details can be found at `https://github.com/google/gnxi`.

- `openconfig-gnmi-gateway` is a library that can be used for high-availability streaming to collect network data with multiple clients. Details can be found at `https://github.com/openconfig/gnmi-gateway`.

- `openconfig-gnmic` is a CLI tool written in Go that you can use to test gNMI capabilities. The CLI implements all gNMI client capabilities. More details can be found at `https://github.com/openconfig/gnmic`.

So far, we have covered the major and most popular libraries used to access network devices via several different methods. Further discussion on this topic can be found in open chat communities such as Slack groups. Examples include `https://devopschat.slack.com/` and `https://alldaydevops.slack.com/`.

Summary

In this chapter, we dove deeper into Python and Go runtime behavior, investigated how libraries are added to both languages, and we saw a few examples of network libraries to use when accessing the network devices.

This chapter provided sufficient information to help you differentiate how Python and Go run and how they can be used with standard and third-party libraries. Now, you should be able to choose a proper language for your network automation based on the requirements of performance, security, maintainability, and reliability. You should also be able to choose a proper method and a library to access your network devices, either for configuration purposes or to collect network data.

In the next chapter, we are going to touch on how to handle errors in Go and Python, and how we can write code to handle exceptions properly in our network automation.

Error Handling and Logging

We described how Python and Go run and how they access the network in the previous chapter; however, we missed two important points when building our network automation solution: how we report program execution events and how we handle errors.

These two topics are not as easy as they seem, and they are, most of the time, implemented in the system poorly. Some network developers might not do it properly because of lack of knowledge, but there are also some developers that don't do it properly because of time constraints and extra time needed for coding.

But are these activities really important? Let's examine these in this chapter. First, let's investigate how and why we handle errors and then why and how we do event logging.

Here are the topics we are going to cover in this chapter:

- Writing code for error handling
- Logging events
- Adding logging to your code

After reading this chapter, you will be able to add effectively code to handle errors and log events in your network development.

Technical requirements

The source code described in this chapter is stored in the GitHub repository at `https://github.com/PacktPublishing/Network-Programming-and-Automation-Essentials/tree/main/Chapter07`.

Writing code for error handling

To see how important handling errors is, we have to think of our system as a whole, including inputs and output. Our code by itself might never experience an error; however, when integrated with other systems, it might cause unpredictable outputs, or might just crash and stop working.

Therefore, handling errors is important to cope with the unpredictability of inputs and protect your code to avoid wrong outputs or crashes. But how do we do that?

First, we need to identify our inputs, and then we create a series of different combinations of values that are sent to our inputs. The behavior of these input combinations are then evaluated by running our code. For a function, we do that by adding unit tests, as we discussed in *Chapter 5*. For the system, we add integration tests and end-to-end testing. Additional techniques were also discussed in *Chapter 5*.

But what is the correct way to write code for handling errors? It will depend on the language.

Writing code to handle errors is quite different in Go compared to Python. Let's see how we do this effectively in Go then in Python.

Adding error handling in Go

The design of the Go language requires explicitly checking errors when they occur, which is different from throwing exceptions and then catching them like in Python. In Go, errors are just values returned by functions, which makes Go coding a bit more verbose and perhaps more repetitive. With Python, you would not need to check errors, as an exception would be raised, but in Go, you have to check the errors. But on the other hand, Go error handling is much simpler compared to Python.

Errors in Go are created by using an `error` type interface, as follows:

```
type error interface {
    Error() string
}
```

As you can see in the preceding code, the error implementation in Go is quite simple by using a method called `Error()` that returns an error message as a string.

The correct way to construct an error in your code is by using either the `errors` or `fmt` standard libraries

The following are two examples of using each of them for a division-by-zero function.

Using the `errors` library looks as follows:

```
func divide(q int, d int) (int, error) {
    if d == 0 {
        return 0, errors.New("division by zero not valid")
```

```
    }
    return q / d, nil
}
```

Using the `fmt` library looks as follows:

```
func divide(q int, d int) (int, error) {
    if d == 0 {
        return 0, fmt.Errorf("divided by zero not valid")
    }
    return q / d, nil
}
```

The preceding two examples produce the same results. For instance, if you call any of the two functions with `fmt.Println(divide(10, 0))`, it will print the following output: `0 divided by zero not valid`.

The major difference between `fmt.Errorf` and `errors.New` is the possibility of formatting the string and adding values. Another point is that `errors.New` is faster because it does not invoke the formatter.

If you are looking to create custom errors, stack traces, and more advanced error features, consider using the `errors` library or third-party libraries such as the popular `pkg/errors` (`https://pkg.go.dev/github.com/pkg/errors`) or `golang.org/x/xerrors` (`https://pkg.go.dev/golang.org/x/xerrors`).

Let's focus now on the best practices when writing code for error handling in Go. The following are the top best practices.

Return errors last and values to 0

When creating a function that returns several values, the error should be placed as the last argument returned. When returning values with errors, use 0 when it is a number and `empty string` for strings, as in the following example:

```
func findNameCount(text string) (string, int, error) {
    if len(text) < 5 {
        return "", 0, fmt.Errorf("text too small")
    }
    . . .
}
```

In the preceding example, the value of the string returned is empty and the value of `int` returned is 0. These values are just suggestions, because when an error is returned, the calling statement will first check whether there is an error before assigning the returned variables. Therefore, the values returned with an error are irrelevant.

Add only information that the caller does not have

When creating the error message, do not add information that is already known by the caller of your function. The following example illustrates this problem:

```
func divide(q int, d int) (int, error) {
    if d == 0 {
        return 0, fmt.Errorf("%d can't be divided by zero", q)
    }
    return q / d, nil
}
```

As you can see in the preceding example, the value of q was returned in the error message. But that is not necessary, because the caller of the `divide` function has this value.

When creating errors to return in your function, do not include any arguments passed in the function, as this is known by the caller. This can lead to information duplication.

Use lowercase and do not end with punctuation

Always use lowercase as your error message will be concatenated with other messages when returning. Most of the time, you should also not use any punctuation, because the error messages will probably be linked together, and punctuation will end up looking odd in the middle of the error message.

The exception of the lowercase rule is when you are referring to the names of functions or methods that already have capital letters.

Add a colon to the error message

A colon (:) is used whenever you want to add information from another error message of a call made inside your code. Let's use the following code as an example:

```
func connect(host string, conf ssh.ClientConfig) error {
    conn, err := ssh.Dial("tcp", host+":22", conf)
    if err != nil {
        return fmt.Errorf("ssh.Dial: %v", err)
    }
    . . .
```

In the preceding example, the connect function encapsulates a call to ssh.Dial. We can add the context of the error to the error message by indicating which call generated the error by adding the name of the call or some information about ssh.Dial, using a colon to separate it if necessary. Note that the config and host arguments are known by the caller of the connect function and therefore should not be added to the error message.

Use defer, panic, and recover

Go has important mechanisms to control how the program flows during errors. That is mostly useful when using goroutines, because one might cause the program to crash, and you might need to be extra careful to avoid unclosed software pipes and software caches; and avoid unfreed memory and unclosed file descriptors

defer

Go defer is used to push the execution to a list that is only executed after the surrounding function has returned or when it crashes. The main intention of defer is to perform a cleanup. Consider the following example of a function that copies data from one file to another and then removes it:

```go
func moveFile(srcFile, dstFile string) error {
    src, err := os.Open(srcFile)
    if err != nil {
        return fmt.Errorf("os.Open: %v", err)
    }

    dst, err := os.Create(dstFile)
    if err != nil {
        return fmt.Errorf("os.Create: %v", err)
    }

    _, err = io.Copy(dst, src)
    if err != nil {
        return fmt.Errorf("io.Copy: %v", err)
    }
    dst.Close()
    src.Close()

    err = os.Remove(srcFile)
    if err != nil {
        return fmt.Errorf("os.Remove: %v", err)
```

```
    }
    return nil
}
```

In the preceding example, if an error occurs in `os.Create`, the function would return before calling `src.Close()`, which means the file has not been closed properly.

The way to avoid having to add the `close` statement repetitively throughout the code is to use `defer`, like the following example:

```
func moveFile(srcFile, dstFile string) error {
    src, err := os.Open(srcFile)
    if err != nil {
        return fmt.Errorf("os.Open: %v", err)
    }
    defer src.Close()

    dst, err := os.Create(dstFile)
    if err != nil {
        return fmt.Errorf("os.Create: %v", err)
    }
    defer dst.Close()

    _, err = io.Copy(dst, src)
    if err != nil {
        return fmt.Errorf("io.Copy: %v", err)
    }

    err = os.Remove(srcFile)
    if err != nil {
        return fmt.Errorf("os.Remove: %v", err)
    }
    return nil
}
```

As you can see in the preceding example, `defer` is used just after a successful `os.Open` and after a successful `os.Create`. Therefore, if there is an error, or if the function ends, it will invoke `dst.Close()` first and then `src.Close()` in a reverse order, like in a **Last In, First Out (LIFO)** queue.

Let's see how to use panic now.

panic

When writing code, if you don't want to handle an error, you can use panic to indicate that you want to stop immediately. In Go, panic can be called in your program by explicitly writing it, but it is also called automatically during runtime if an error occurs. The following is a list of the major runtime errors that can occur:

- Out-of-bounds memory access, including arrays
- Assertion of a wrong type
- Attempting to call a function using a variable with a nil pointer
- Sending data to a closed channel or file descriptor
- Division by zero

Therefore, panic is only used in your code when you are not planning to handle the error or are dealing with errors that are not yet understood.

It is important to note that before exiting the function and passing the panic message, the program will still run all defer statements that were stacked earlier in the function.

Here is an example of using panic to exit the program after receiving a negative value as an argument:

```go
import (
    "fmt"
    "math"
)

func squareRoot(value float64) float64 {
    if value < 0 {
        panic("negative values are not allowed")
    }

    return math.Sqrt(value)
}

func main() {
    fmt.Println(squareRoot(-2))
    fmt.Println("done")
}
```

Let's run this program and check the output:

```
$ go run panic-example.go
panic: negative values are not allowed

goroutine 1 [running]:
main.squareRoot(...)
    Dev/Chapter07/Go/panic-example.go:10
main.main()
    Dev/Chapter07/Go/panic-example.go:17 +0x45
exit status 2
```

Note that the output does not print done, because panic is called inside the squareRoot function, before the instruction to print.

Say we add defer to the function as follows:

```
func squareRoot(value float64) float64 {
    defer fmt.Println("ending the function")
    if value < 0 {
        panic("negative values are not allowed")
    }

    return math.Sqrt(value)
}
```

The output will be as follows:

```
$ go run panic-example.go
ending the function
panic: negative values are not allowed

goroutine 1 [running]:
main.squareRoot(...)
    Dev/Chapter07/Go/panic-example.go:10
main.main()
    Dev/Chapter07/Go/panic-example.go:17 +0x45
exit status 2
```

Note that the ending the function print statement was placed before sending the panic message. That is because, as we explained, the defer stack is executed before returning the function by panic.

Let's now see how we can use recover.

recover

In Go, recover is the last piece of error flow control necessary to handle errors. It is used to handle a panic situation and regain control. It should only be used inside a defer function call. In a normal call, recover will return a nil value, but in the panic situation, it will return the value given to panic.

As an example, let's consider the following program:

```
import "fmt"

func divide(q, d int) int {
    fmt.Println("Dividing it now")


    return q / d
}

func main() {
    fmt.Println("the division is:", divide(4, 0))
}
```

If you run the preceding program, you will get the following panic message:

```
$ go run division-by-zero-panic.go
Dividing it now
panic: runtime error: integer divide by zero

goroutine 1 [running]:
main.divide(...)
    Dev/Chapter07/Go/division-by-zero-panic.go:7
main.main()
    Dev/Chapter07/Go/division-by-zero-panic.go:11 +0x85
exit status 2
```

As you can see from this output, you have no control of the panic situation. It will basically crash the program without the possibility to handle the error properly. This is undesirable in most production software, especially using multiple goroutines.

Therefore, to properly handle the panic situation, you should add a defer function to test whether it is a panic situation by using recover, as in the following example:

```
import "fmt"

func divide(q, d int) int {
    fmt.Println("Dividing it now")

    return q / d
}

func main() {
    defer func() {
        if r := recover(); r != nil {
            fmt.Println("Got a panic:", r)
        }
    }()

    fmt.Println("the division is:", divide(4, 0))
}
```

After adding the defer function as in the preceding example, the output will be as follows:

```
$ go run division-by-zero-panic-recover.go
Dividing it now
Got a panic: runtime error: integer divide by zero
```

As you can see, adding a recover test inside the defer function will allow you to handle unexpected panic situations, avoiding your program crashing unexpectedly without doing a proper cleanup or fixing the error.

Now that we have investigated how to handle Go errors, let's have a look at Python error handling.

Adding error handling in Python

Python handles errors differently than Go. Python does not require your function to return error values. In Python, errors are thrown during runtime and they are called exceptions. To handle exceptions, your code has to catch them properly and avoid raising them.

Catching exceptions

In Python, there are built-in exceptions that are raised by many runtime errors. The list of built-in exceptions is quite long and can be found here: https://docs.python.org/3/library/exceptions. html. As an example, the division-by-zero error is called a `ZeroDivisionError` exception.

To handle the error, you need to catch the exception and then handle it by using the `try`, `except`, `else`, and `finally` Python statements. To create an example of how to handle the exception for division by zero, let's first run the following program without catching an exception:

```python
def division(q, d):
    return q/d
print(division(1, 0))
```

If you run the preceding program, it will generate the following output:

```
$ python catching-division-by-zero-exception.py
Traceback (most recent call last):
  File "Chapter07/Python/catching-division-by-zero-exception.
py", line 7, in <module>
    print(division(1, 0))
  File "Chapter07/Python/catching-division-by-zero-exception.
py", line 4, in division
    return q/d
ZeroDivisionError: division by zero
```

As you can see, the program crashes and shows the error message as `Traceback` on the screen, with details of where the error occurs and the name of the exception, in this case, `ZeroDivisionError`.

Now, let's update the Python code to catch this exception and handle the error more gracefully, like the following code:

```python
def division(q, d):
    return q/d
try:
    print(division(1, 0))
```

```
except ZeroDivisionError:
    print("Error: We should not divide by zero")
```

Now, if you run the program, it will print the error gracefully without crashing it, as follows:

```
$ python catching-division-by-zero-exception.py
Error: We should not divide by zero
```

So, whenever you think there is a possibility for the function to raise an exception by an error, use the try and except statements, as shown in the preceding example.

In addition to try and except statements, Python also allows using else and finally statements to add more flow control of the error handling. They are not mandatory as the flow can be controlled outside the try/except statement, but they are sometimes useful. The following is the same example of adding the else and finally statements:

```
def division(q, d):
    return q/d


try:
    result = division(10, 1)
except ZeroDivisionError:
    print("Error: We should not divide by zero")
else:
    print("Division succeded, result is:", result)
finally:
    print("done")
```

If you run this program, it will generate the following output:

```
$ python catch-else-finally-division-by-zero.py
Division succeded, result is: 10.0
done
```

Note that the else statement is only executed if no exception was raised in the try clause. The finally statement is always executed, regardless of whether an exception was raised or not in the try clause.

Now that we have seen how to catch an exception in Python, let's discuss how to choose the exception we want to catch.

Choosing more specific exceptions

In Python, exceptions are hierarchical and always start with the exception called `BaseException`. As an example, division by zero exhibits the following hierarchy:

```
BaseException -> Exception -> ArithmeticError->
ZeroDivisionError
```

The exceptions hierarchy is quite useful, because your code can catch either a higher-level exception or a more specific one. With division by zero, you could catch the `ArithmeticError` exception instead of `ZeroDivisionError`. However, it's a good practice sometimes to catch the more specific exception instead of a higher-level one.

More specific exceptions are more desirable to be caught inside functions and libraries, because if you catch generic exceptions inside the function, you might mask the problem when another part of the code calls your function. So, it will depend on where you are catching and how you are handling it.

We now have a good idea how to handle errors in Go and Python. Let's discuss how to add logging to our code.

Logging events

In computer software, logging is a well-known technique used to help troubleshoot a problem, record milestones, understand behavior, retrieve information, and check historical events, among other useful actions. Despite these advantages, not many developers add proper logging to their code. In fact, some developers do nothing and add logging only when the program has problems and needs debugging.

In network automation, logging is even more important, because network elements are normally distributed and rely heavily on logging to be able to be audited in case of a problem or if an improvement is needed. Adding logging to your code is a good practice that will be appreciated by several levels of engineering, such as network operators, network planners, network security, and network designers, among others.

But one important point here that must be observed is that time synchronization between network elements is mandatory to allow logging to be useful. Protocols such as the **Network Time Protocol (NTP)** or **Precision Time Protocol (PTP)** must be used throughout the network.

A good practice to use logging is to use the Unix logging reference called `syslog`, firstly published as an informational RFC in RFC3164 and then as a standard document in RFC5424 (`https://www.rfc-editor.org/rfc/rfc3164`).

For our network automation code, we do not need to follow all the details in the `syslog` protocol standard, but we are going to use it as a guide for logging useful information based on severity level.

Let's now talk about some levels of information we want when logging events.

Severity levels

The RFC5424 `syslog` protocol has defined eight levels of severity, which are described in *Section 6.2.1* of RFC5424 (`https://www.rfc-editor.org/rfc/rfc5424#section-6.2.1`). They are mentioned in the following list with a brief explanation of what kind of information message is intended to be added to each one:

- `Emergency`: The system is not operational and there is no possible recovery.
- `Alert`: Immediate attention is required.
- `Critical`: Something bad is happening and quick attention is required to fix it.
- `Error`: Failure is occurring but does not need urgent attention.
- `Warning or Warn`: Indicates that something is wrong and might cause an error in the future, such as software not updated.
- `Notice`: An important milestone that has been reached and might indicate a future warning, such as configuration not saved or resource utilization limit not set.
- `Informational or Info`: Normal operational milestone messages. Used later for audit and investigation.
- `Debug`: Used by developers to troubleshoot a problem or to investigate possible improvements.

Although these eight levels are defined in the `syslog` protocol, they are quite ambiguous and open to different interpretations. For instance, `Alert` and `Emergency` might be different for different developers when writing code, as is the case with other levels, such as `Notice` and `Informational`. Therefore, some network developers prefer to use less levels with easier interpretation. The number will depend on how the network operates, but varies between three and five levels. For Go and Python, the number of levels will depend on the library you are using to create the log messages. Some might have more levels than others.

Now, let's investigate how to add log events to your code using Go and Python.

Adding logging to your code

Adding event logging in your code will be different in Go and in Python and will vary depending on the library used in your code. But the idea for both languages is to divide the information into severity levels, as done in `syslog`.

The severity log levels will also vary depending on the library used. Python and Go have standard log libraries, but you are also able to use third-party libraries to log events in both languages.

One important point here is that when writing the code, you will decide whether there is a need to add a logging event line. The line of logging added must carry some information that will signal to the program which level of severity the message is. Therefore, important messages such as failures will have priority over less-important messages, such as debugging. Ideally, the decision about which level of logging should be exposed is normally made by adding input arguments to your program that allow setting the log level. So, if you are running the program for debugging, it will generate much more information compared to a normal operation.

Let's now see how we can add logging events to our code in Go and then we check how to do it in Python.

Adding event logging in Go

The Go language has a standard library for logging that comes with the Go installation, but it is quite limited. For more advanced logging capabilities in Go, you might want to use third-party libraries for logging.

Let's see how we can use the standard library and then check other popular third-party libraries.

Using standard Go logging

The standard logging library in Go can be imported using `log` in the `import` statement. By default, Go standard logging does not provide any severity levels, but it has some helper functions that can help create logs. The helper functions are listed here:

- `Print`, `Printf`, and `Println`: These functions print the message passed to them in the terminal using `stderr`
- `Panic`, `Panicf`, and `Panicln`: These are like `Print`, but they call `Panic` after printing the log message
- `Fatal`, `Fatalf`, and `Fatalln`: These also work like `Print`, but they call `os.Exit(1)` after printing the log message

The following is a simple example of using the standard Go logging library:

```
import (
    "log"
    "os/user"
)

func main() {
    user, err := user.Current()
    if err != nil {
        log.Fatalf("Failed with error: %v", err)
```

```
    }

    log.Printf("Current user is %s", user.Username)
}
```

Running this program will print the following output without errors:

```
% go run standard-logging.go
2022/11/08 18:53:24 Current user is claus
```

If for any reason it is not possible to retrieve the current user, it will call `Fatalf`, which will call `os.Exit(1)` after printing the failed message.

Now, let's show a more complex example of how to create severity levels using the standard logging library and saving it to a file:

```
import (
    "log"
    "os"
)

var criticalLog, errorLog, warnLog, infoLog, debugLog *log.
Logger

func init() {
    file, err := os.Create("log-file.txt")
    if err != nil {
        log.Fatal(err)
    }
    flags := log.Ldate | log.Ltime
    criticalLog = log.New(file, "CRITICAL: ", flags)
    errorLog = log.New(file, "ERROR: ", flags)
    warnLog = log.New(file, "WARNING: ", flags)
    infoLog = log.New(file, "INFO: ", flags)
    debugLog = log.New(file, "DEBUG: ", flags)
}

func main() {
    infoLog.Print("That is a milestone")
```

```
        errorLog.Print("Got an error here")
        debugLog.Print("Extra information for a debug")
        warnLog.Print("You should be warned about this")
}
```

In the preceding example, we created five levels of severity that can be used to write to a file as required. Note that in Go, the `init()` function is executed before the `main()` function. If you want to use these log definitions within other packages, remember to use capitalization of the variables; otherwise, the variables will be local to this package; for example, `errorLog` should be `ErrorLog`.

Also, if you want to set a log level to avoid `Debug` or `Info` messages, you will have to pass an argument to your program and suppress lower levels of severity depending on the level set. Using the Go standard logging library, you will have to do that on your own.

Now, let's investigate a third-party logging library that is very popular with Go developers.

Using logrus

Perhaps one of the most popular libraries for logging in Go, `logrus` is a structured logging library with several logging capabilities. `logrus` has seven log levels and is compatible with the standard logging library. By default, the library allows you to set the log level, so it won't create noise if you don't want to see debugging information.

Here is a simple example of using `logrus` and setting the log level to `Error`, which means lower-level logs will not show, such as `Warning`, `Info`, or `Debug`:

```
import (
    log "github.com/sirupsen/logrus"
)

func init() {
    log.SetFormatter(&log.TextFormatter{
        DisableColors: true,
        FullTimestamp: true,
    })
    log.SetLevel(log.ErrorLevel)
}

func main() {
    log.Debug("Debug is suppressed in error level")
    log.Info("This info won't show in error level")
```

```
        log.Error("Got an error here")
}
```

Running the preceding example will only show the following output in the terminal:

```
% go run logrus-logging.go
time="2022-11-09T11:16:48-03:00" level=error msg="Got an error
here"
```

Because the level of severity is set to ErrorLevel, none of the less-significant log messages will be displayed – in the example, the calls for log.Info and log.Debug.

logrus is very flexible and powerful with plenty of examples of usage on the internet. More details on logrus can be found at https://github.com/sirupsen/logrus.

If you want to use more logging libraries in Go, here is a compiled list of third-party log libraries: https://awesome-go.com/logging/.

Now, let's check how we add logging to our code using Python.

Add event logging in Python

In comparison to Go, Python has added many more features to the standard logging library. Despite the better support, the Python community has also developed many third-party logging libraries.

Let's have a look at the standard library and then a popular third-party library for Python.

Using standard logging for Python

The standard logging library comes with five severity levels and an extra level that is used to indicate the level was not set by the logger. Each level has a number associated with it that can be used to interpret the priority level, where lower numbers have lower priority. The levels are CRITICAL (50), ERROR (40), WARNING (30), INFO (20), DEBUG (10), and NOTSET (0).

The NOTSET level is useful when using log hierarchies, allowing a non-root logger to delegate the level to its parent.

The following is an example of using Python standard logging:

```python
import logging

logging.basicConfig(
    filename='file-log.txt',
    level=logging.ERROR,
    format='%(asctime)s.%(msecs)03d %(levelname)s:
```

```
%(message)s',
        datefmt='%Y-%m-%d %H:%M:%S',
)
logging.debug("This won't show, level is set to info")
logging.info("Info is not that important as well")
logging.warning("Warning will not show as well")
logging.error("This is an error")
```

Running the preceding program will produce the following line in the output file called `file-log.txt`:

```
2022-11-09 14:48:52.920 ERROR: This is an error
```

As illustrated in the preceding code, setting the level to `logging.ERROR` will not allow the lower-level log messages to be written in the file. The program just ignores the `logging.debug()`, `logging.info()`, and `logging.warning()` calls.

Another important point is to show how easier it is to use standard logging in Python. The preceding example shows that you just need one call to `logging.basicConfig` to set almost everything you need, from the formatter to the level of severity.

In addition to being easy to use, the Python community has created great tutorials and documentation for the standard logging library. Here are the three main references for the documentation and advanced utilization info:

- `https://docs.python.org/3/library/logging`
- `https://docs.python.org/3/howto/logging.html`
- `https://docs.python.org/3/howto/logging-cookbook.html`

In essence, the Python standard logging library is quite complete, and you won't need to use a third-party library for most of your work. However, there are some nice and interesting features in one popular third-party library for Python called `loguru`. Let's see how to use it.

Using Python loguru

Python `loguru` provides a few more features than the standard Python logging library and has the aim of being easier to use and configure. For instance, using `loguru`, you will be able to set file rotation on the log file, use a more advanced string formatter, and use decorators to catch exceptions on functions, and it is thread and multiprocessing safe.

It also has interesting features that can allow you to add extra information to the logging by using a `patch` method (more on the `patch` method at `https://loguru.readthedocs.io/en/stable/api/logger.html#loguru._logger.Logger.patch`).

The following is a simple example using `loguru`:

```
from loguru import logger

logger.add(
    "file-log-{time}.txt",
    rotation="1 MB",
    colorize=False,
    level="ERROR",
)


logger.debug("That's not going to show")
logger.warning("This will not show")
logger.error("Got an error")
```

Running the preceding example will create a file with the date and time with log messages that will rotate if it reaches 1 MB in size. The output written in the file will look as follows:

```
% cat file-log-2022-11-09_15-53-58_056790.txt
2022-11-09 15:53:58.063 | ERROR    | __main__:<module>:13 - Got
an error
```

More detailed documentation can be found at `https://loguru.readthedocs.io` and the source code at `https://github.com/Delgan/loguru`.

Summary

After reading this chapter, you are probably more aware of why we need to handle errors and why we need to create proper event logging. You also should be more familiar with the differences between how Go and Python handle errors. Furthermore, you saw the differences in how to use standard libraries and third-party libraries for event logging. From now on, your network automation code design will have a special section on logging and error handling.

In the next chapter, we are going to talk about how we can scale our code and how our network automation solution can interact with large networks.

8

Scaling Your Code

Now that we know how to interact with network devices, we should start thinking about building a solution that scales. But why do we need to scale our code? You might be thinking that the answer is obvious and it is just because it will allow your solution to grow easily as your network grows. But scaling is not just about going up but scaling down too. So, scaling your code means that you are going to build a solution that can follow demand easily, saving resources when not required and using more when required.

You should consider adding scaling capabilities adding scaling capabilities to your network automation solution before writing the code. It should be planned during design time and then executed during development time. The scaling capabilities have to be one of the requirements for building your solution. It also should be a clear milestone during implementation and testing.

In this chapter, we are going to check some techniques used today to scale your code up and down effectively. This will allow your solution to adapt easily to follow network growth and, if necessary, easily scale down to save resources.

We are going to cover the following topics in this chapter:

- Dealing with multitasking, threads, and coroutines
- Adding schedulers and job dispatchers
- Using microservices and containers

By the end of this chapter, you should have enough information to choose the best scaling solution for your code.

Technical requirements

The source code described in this chapter is stored in the GitHub repository at `https://github.com/PacktPublishing/Network-Programming-and-Automation-Essentials/tree/main/Chapter08`.

Dealing with multitasking, threads, and coroutines

Multitasking, as the name suggests, is the capability of doing several tasks at the same time. In computers, a task is also known as a job or a process and there are different techniques for running tasks at the same time.

The capability to run code at the same time allows your system to scale up and down whenever necessary. If you have to communicate with more network devices, just run more code in parallel; if you need fewer devices, just run less code. That will enable your system to scale up and down.

But running code in parallel will have an impact on the available machine resources, and some of these resources will be limited by how your code is consuming them. For instance, if your code is using the network interface to download files, and running one single line of code is already consuming 50 Mbps of the network interface (which is 50% of the 100 Mbps interface), it is not advised to run multiple lines of code in parallel to increase the speed, as the limitation is on the network interface and not in the CPU.

Other factors are also important to be considered when running code in parallel, that is, the other shared resources besides the network, such as the CPU, disk, and memory. In some cases, bottlenecks in the disk might cause more limitations for code parallelism than the CPU, especially using disks mounted over the network. In other cases, a large program consuming lots of memory would block the execution of any other program running in parallel because of a lack of free memory. Therefore, the resources that your process will touch and how they interact will have an impact on how much parallelism is possible.

One thing we should clarify here is the term **I/O**, which is an acronym for computer **input/output**. I/O is used to designate any communication between the CPU of the machine and the external world, such as accessing the disk, writing to memory, or sending data to the network. If your code requires lots of external access and it is, most of the time, waiting to receive a response from external communication, we normally say the code is **I/O bound**. An example of slow I/O can be found when accessing remote networks and, in some cases, remote disks. On the other hand, if your code requires more CPU computation than I/O, we normally say the code is **CPU bound**. Most network automation systems will be I/O bound because of network device access.

Let's now investigate a few techniques to run code at the same time in Go and Python.

Multiprocessing

In computers, when a program is loaded in memory to run, it's called a process. The program can be either a script or a binary file, but it is normally represented by one single file. This file will be loaded into memory and it is seen by the operating system as a process. The capability of running multiple processes at the same time is called multiprocessing, and it is normally managed by the operating system.

The number of CPUs of the hardware where the processes are running is irrelevant to the multiprocessing capability. The operating system is responsible for allocating the CPU time for all processes that are in memory and ready to run. However, as the number of CPUs, speed of the CPU, and memory are limited, the number of processes that can run at the same time will also be limited. Normally, it depends on the size of the process and how much CPU it consumes.

In most computer languages, multiprocessing is implemented using the fork() system call implemented by the operating system to create a complete copy of the currently running process.

Let's investigate how we can use multiprocessing in Go and Python.

Multiprocessing in Python

In Python, multiprocessing is accomplished by the standard library called multiprocessing. Full documentation on Python multiprocessing can be found at docs.python.org/3/library/multiprocessing.

In the first example, we are going to use the operating system program called ping to target one network node. Then, we are going to make it parallel for multiple targets.

The following is an example for one target network node:

```python
import subprocess

TARGET = "yahoo.com"

command = ["ping", "-c", "1", TARGET]
response = subprocess.call(
    command,
    stdout=subprocess.DEVNULL,
)
if response == 0:
    print(TARGET, "OK")
else:
    print(TARGET, "FAILED")
```

It is important to note that calling `ping` from Python is not efficient. It will cause more overhead because Python will have to invoke an external program that resides in the filesystem. In order to make the example more efficient, we need to use the ICMP `echo request` and receive an ICMP `echo reply` from the Python network sockets, instead of invoking an external program such as `ping`. One solution is to use the Python third-party library called `pythonping` (https://pypi.org/project/pythonping/). But there is one caveat: the `ping` program has `setuid` to allow ICMP packets to be sent by a non-privileged user. Thus, in order to run with `pythonping`, you need admin/root privileges (accomplished in Linux using `sudo`).

The following is the same example using `pythonping` for one target network node:

```
import pythonping

TARGET = "yahoo.com"

response = pythonping.ping(TARGET, count=1)
if response.success:
    print(TARGET, "OK")
else:
    print(TARGET, "FAILED")
```

Running this program should generate the following output:

```
% sudo python3 single-pyping-example.py
yahoo.com OK
```

If you want to send ICMP requests to multiple targets, you will have to send them sequentially one after the other. However, a better solution would be to run them in parallel using the `multiprocessing` Python library. The following is an example of four targets using `multiprocessing`:

```
from pythonping import ping
from multiprocessing import Process

TARGETS = ["yahoo.com", "google.com", "cisco.com", "cern.ch"]

def myping(host):
    response = ping(host, count=1)
    if response.success:
        print("%s OK, latency is %.2fms" % (host, response.
rtt_avg_ms))
```

```
        else:
            print(host, "FAILED")

def main():
    for host in TARGETS:
        Process(target=myping, args=(host,)).start()

if __name__ == "__main__":
    main()
```

If you run the preceding program, you should get an output similar to the following:

```
% sudo python3 multiple-pyping-example.py
google.com OK, latency is 45.31ms
yahoo.com OK, latency is 192.17ms
cisco.com OK, latency is 195.44ms
cern.ch OK, latency is 272.97ms
```

Note that the response of each target does not depend on the response of others. Therefore, the output should always be in order from low latency to high latency. In the preceding example, google.com finished first, showing a latency of just 45.31 ms.

> **Important note**
>
> It is important to call multiprocessing inside the main() function, or a function that is called from main(). Also, make sure main() can be safely imported by a Python interpreter (use __name__). You can find more details on why at https://docs.python.org/3/library/multiprocessing.html#multiprocessing-programming.

Python has additional methods to invoke code parallelism besides the preceding example using Process(), called multiprocessing.Pool() and multiprocessing.Queue(). The Pool() class is used to instantiate a pool of workers that can do a job without the need to communicate with each other. The Queue() class is used when communication between processes is required. More on that can be found at https://docs.python.org/3/library/multiprocessing.html.

Let's see how we can use multiprocessing in Go.

Multiprocessing in Go

To create processes from a program, you need to create a copy of the data of the current running program to a new process. That is what Python's `multiprocessing` does. However, Go implements parallelism very differently. Go was designed to work with routines similar to coroutines, they are called goroutines, which manage parallelism at runtime. As goroutines are much more efficient, there is no need to implement multiprocessing natively in Go.

Note that using the `exec` library, by calling `exec.Command()` and then `Cmd.Start()` and `Cmd.Wait()`, will allow you to create multiple processes at the same time, but it is a call to the operating system to execute an external program. Therefore, it is not considered native multiprocessing and is not efficient.

For these reasons, we don't have an example of multi-processing in Go.

Let's see now how we do multithreading.

Multithreading

In computer languages, a thread is a smaller part of a process, which can have one or multiple threads. The memory is shared between the threads in the same process, in contrast with a process that does not share memory with another process. Therefore, a thread is known as a lightweight process because it requires less memory, and communication between threads within a process is faster. In consequence, spawning new threads is much faster in comparison with new processes.

A CPU with multithreading capability is a CPU that has the ability to run multiple threads in a single core by providing instruction-level parallelism or thread-level parallelism. This capability is also known as **Simultaneous Multithreading (SMT)**.

One example of SMT is the Intel CPU i9-10900K, which has 10 cores and the capability to run 2 threads at the same time per core, which allows up to 20 simultaneous threads. Intel has created a trademark name for SMT, which they call **hyper-threading**. Normally, AMD and Intel x86 CPU architectures can run up to two threads per core.

In contrast, the Oracle SPARC M8 processor has 32 cores that can run 8 threads each, allowing a staggering number of 256 simultaneous threads. More on this amazing CPU can be found at `https://www.oracle.com/us/products/servers-storage/sparc-m8-processor-ds-3864282.pdf`.

But for the CPU to perform its best using threads, two other requirements are necessary, an operating system that allows CPU-level multithreading and a computer language that allows the creation of simultaneous threads.

Let's see how we can use multithreading in Python.

Multithreading in Python

Multithreading is the Achilles' heel of Python. The main reason is that the Python interpreter called CPython (discussed in *Chapter 6*) uses a **Global Interpreter Lock** (**GIL**) to make it thread-safe. This has a consequence of not allowing Python code to run multiple threads at the same time in a multithread CPU. GIL also adds overhead and using multithreading might cause the program to run slower in comparison with multiprocessing when more CPU work is required.

Therefore, in Python, multithreading is not recommended for programs that are CPU bound. For network and other I/O-bound programs, multithreading might be faster to spawn and easier to communicate with and save runtime memory. But it is important to note that only one thread will run at a time using the CPython interpreter, so if you require true parallelism, use the `multiprocessing` library instead.

In Python, the standard library offers multithreading by using the library called `threading`. So, let's create one example using multithreading in Python by taking the same targets for ICMP tests used in the code example in the previous section. The following is the same example using ICMP but using threading:

```python
from pythonping import ping
import threading

TARGETS = ["yahoo.com", "google.com", "cisco.com", "cern.ch"]

class myPing(threading.Thread):
    def __init__(self, host):
        threading.Thread.__init__(self)
        self.host = host
    def run(self):
        response = ping(self.host)
        if response.success:
            print("%s OK, latency is %.2fms" % (self.host,
response.rtt_avg_ms))
        else:
            print(self.host, "FAILED")

def main():
    for host in TARGETS:
        myPing(host).start()
```

```
if __name__ == "__main__":
    main()
```

The output of running the preceding program will look as follows:

```
% sudo python3 threads-pyping-example.py
google.com OK, latency is 36.21ms
yahoo.com OK, latency is 136.16ms
cisco.com OK, latency is 144.67ms
cern.ch OK, latency is 215.81ms
```

As you can see, the output is quite similar using the threading and multiprocessing libraries, but which one runs faster?

Let's now run a test program to compare the speed of using threading and multiprocessing for the ICMP tests. The source code of this program is included in the GitHub repository for this chapter. The name of the program is performance-thread-process-example.py.

Here is the output of this program running for 10, 20, 50, and 100 ICMP probes:

```
% sudo python3 performance-thread-process-example.py 10
Multi-threading test --- duration 0.015 seconds
Multi-processing test--- duration 0.193 seconds

% sudo python3 performance-thread-process-example.py 20
Multi-threading test --- duration 0.030 seconds
Multi-processing test--- duration 0.315 seconds
% sudo python3 performance-thread-process-example.py 50
Multi-threading test --- duration 2.095 seconds
Multi-processing test--- duration 0.765 seconds
% sudo python3 performance-thread-process-example.py 100
Multi-threading test --- duration 2.273 seconds
Multi-processing test--- duration 1.507 seconds
```

As shown in the preceding output, running multithreading in Python for a certain number of threads might be faster. However, as we get close to the number 50, it becomes less effective and runs much slower. It is important to notice that this will depend on where you are running your code. The Python interpreter running on Windows is different from in Linux or even in macOS, but the general idea is the same: more threads mean more overhead for the GIL.

The recommendation is not to use Python multithreading unless you are spawning a small number of threads and are not CPU bound.

> **Important note**
>
> Because of the CPython GIL, it is not possible to run parallel threads in Python. Therefore, if your program is CPU bound and requires CPU parallelism, the way to go is to use the `multiprocessing` library instead of the `threading` library. More details can be found at `docs.python.org/3/library/threading`.

But if you still want to use Python with multithreading, there are other Python interpreters that might provide some capability. One example is **PyPy-STM** (PyPy was introduced and explained in *Chapter 6*). However, to allow the usage of PyPy-STM, you will have to rewrite your code and not use the default `threading` module. With PyPy-STM, it is possible for simultaneous threads to run, but you will have to use the `transaction` module, specifically the `TransactionQueue` class. More on multithreading using PyPy-STM can be found at `doc.pypy.org/en/latest/stm.html#user-guide`.

Now, let's see how we can do multithreading in Go.

Multithreading in Go

Writing code that scales in Go does not require the creation of threads or processes. Go implements parallelism very efficiently using goroutines, which are presented as threads to the operating system by the Go runtime. Goroutines will be explained in more detail in the following section, which talks about coroutines.

We will also see how we can run multiple lines of code at the same time using coroutines.

Coroutines

The term *coroutine* was coined back in 1958 by Melvin Conway and Joel Erdwinn. Then, the idea was officially introduced in a paper published in the *ACM* magazine in 1963.

Despite being very old, the adoption of the term came later with some modern computer languages. Coroutines are essentially code that can be suspended. The concept is like a thread (in multithreading), because it is a small part of the code, and has local variables and its own stack. But the main difference between threads and coroutines in a multitasking system is threads can run in parallel and coroutines are collaborative. Some like to describe the difference as the same as between task concurrency and task parallelism.

Here is the definition taken from *Oracle Multithreaded Programming Guide*:

In a multithreaded process on a single processor, the processor can switch execution resources between threads, resulting in concurrent execution. Concurrency indicates that more than one thread is making progress, but the threads are not actually running simultaneously. The switching between threads happens quickly enough that the threads might appear to run simultaneously. In the same multithreaded process in a shared-memory multiprocessor environment, each thread in the process can run concurrently on a separate processor, resulting in parallel execution, which is true simultaneous execution.

The source can be found at `https://docs.oracle.com/cd/E36784_01/html/E36868/mtintro-6.html`.

So, let's now check how we can use coroutines in Python and then in Go.

Adding coroutines in Python

Python has recently added coroutines to the standard library. They are part of the module called `asyncio`. Because of that, you won't find this capability for older versions of Python; you need at least Python version 3.7.

But when do we use coroutines in Python? The best fit is when you require lots of parallel tasks that are I/O bound, such as a network. For CPU-bound applications, it is always recommended to use `multiprocessing` instead.

In comparison to `threading`, `asyncio` is more useful for our network automation work, because it is I/O bound and scales up more than using `threading`. In addition, it is even lighter than threads and processes.

Let's then create the same ICMP probe test using coroutines in Python. The following is an example of the code for the same network targets used in previous examples (you can find this code in `Chapter08/Python/asyncio-example.py` in the GitHub repo of the book):

```python
from pythonping import ping
import asyncio

TARGETS = ["yahoo.com", "google.com", "cisco.com", "cern.ch"]

async def myping(host):
    response = ping(host)
    if response.success:
        print("%s OK, latency is %.3fms" % (host, response.rtt_avg_ms))
    else:
        print(host, "FAILED")
```

```
async def main():
    coroutines = []
    for target in TARGETS:
        coroutines.append(
            asyncio.ensure_future(myping(target)))
    for coroutine in coroutines:
        await coroutine

if __name__ == "__main__":
    asyncio.run(main())
```

Running the preceding program example will generate the following output:

```
% sudo python3 asyncio-example.py
yahoo.com OK, latency is 192.75ms
google.com OK, latency is 29.93ms
cisco.com OK, latency is 162.89ms
cern.ch OK, latency is 339.76ms
```

Note that now, the first ping reply to be printed is not actually the one that has the least latency, which shows the program is running sequentially, following the order of the TARGETS variable in the loop. That means the asyncio coroutines are not being suspended to allow others to run when they are blocked. Therefore, this is not a good example of using coroutines if we want to scale up. This is because the library used in the example is pythonping, which is not asyncio compatible and is not suspending the coroutine when it is waiting for the network ICMP response.

We added this example to show how bad it is to use asyncio with coroutines that have code that is incompatible with asyncio. To fix this issue, let's now use a third-party library for the ICMP probe that is compatible with asyncio, which is called aioping.

The following code only shows the change on the import to add aioping instead of pythonping and the change on the myping() function, where we added an await statement before the ping() function. The other difference is that aioping works with the exception called TimeoutError to detect a non-response of an ICMP request:

```
from aioping import ping

async def myping(host):
    try:
        delay = await ping(host)
        print("%s OK, latency is %.3f ms" % (host, delay *
```

```
1000))
    except TimeoutError:
        print(host, "FAILED")
```

The complete program with the fixes shown previously can be found in the GitHub repository of this book at `Chapter08/Python/asyncio-example-fixed.py`.

If you run this code now with the fix, it should show something like the following output:

```
% sudo python3 asyncio-example-fixed.py
google.com OK, latency is 40.175 ms
cisco.com OK, latency is 170.222 ms
yahoo.com OK, latency is 181.696 ms
cern.ch OK, latency is 281.662 ms
```

Note that, now, the output is based on how fast the targets answer the ICMP request and the output does not follow the `TARGETS` list order like in the previous example.

The important difference in the preceding code is the usage of `await` before `ping`, which indicates to the Python `asyncio` module that the coroutine may stop and allow another coroutine to run.

Now, you may be wondering whether you could, instead of using the new library, `aioping`, just add `await` to the previous example in front of the `ping` statement in the `pythonping` library. But that will not work and will generate the following exception:

```
TypeError: object ResponseList can't be used in 'await'
expression
```

That is because the `pythonping` library is not compatible with the `asyncio` module.

Use `asyncio` whenever you need to have lots of tasks running because it is very cheap to use coroutines as a task, much faster and lighter than processes and threads. However, it requires that your application be I/O bound to take advantage of the concurrency of the coroutines. Access to network devices is a good example of a slow I/O-bound application and may be a perfect fit for our network automation cases.

> **Important note**
>
> To allow efficient use of coroutines in Python, you have to make sure that the coroutine is suspending execution when there is a wait in I/O (such as a network) to allow other coroutines to run. This is normally indicated by the `asyncio` statement called `await`. Indeed, using the third-party library in your coroutine needs to be compatible with `asyncio`. As the `asyncio` module is quite new, there are not many third-party libraries that are compatible with `asyncio`. Without this compatibility, your code will run coroutines sequentially instead of concurrently, and using `asyncio` will not be a good idea.

Let's see how coroutines can be used in Go.

Coroutines in Go

The Go language is special and shines best when it requires code to scale up with performance, and that is accomplished in Go using goroutines.

Goroutines are not the same as coroutines, because they can run like threads in parallel. But they are not like threads either, because they are much smaller (starting with only 8 KB for Go version 1.4) and use channels for communication. This may be confusing at first, but I promise you that goroutines are not difficult to understand and use. Indeed, they are easier to understand and use compared to coroutines in Python.

Since Go version 1.14, goroutines are implemented using asynchronously preemptible scheduling. That means the tasks are no longer in the control of the developer and are entirely managed by Go's runtime (you can find details at `https://go.dev/doc/go1.14#runtime`). Go's runtime is responsible for presenting to the operating system the threads that are going to run, which can run simultaneously in some cases.

Go's runtime is responsible for creating and destroying the threads that correspond to a goroutine. These operations would be much heavier when implemented by the operating system using a native multithreading language, but in Go, they are light as Go's runtime maintains a pool of threads for the goroutines. The fact that Go's runtime controls the mapping between goroutines and threads makes the operating system completely unaware of goroutines.

In summary, Go doesn't use coroutines, but instead uses goroutines, which are not the same and are more like a blend between coroutines and threads, with better performance than the two.

Let's now go through a simple example of an ICMP probe using goroutines:

```go
import (
    "fmt"
    "time"
    "github.com/go-ping/ping"
)


func myPing(host string) {
    p, err := ping.NewPinger(host)
    if err != nil {
        panic(err)
    }
    p.Count = 1
```

```
    p.SetPrivileged(true)
    if err = p.Run(); err != nil {
        panic(err)
    }
    stats := p.Statistics()
    fmt.Println(host, "OK, latency is", stats.AvgRtt)
}


func main() {
    targets := []string{"yahoo.com", "google.com", "cisco.com",
"cern.ch"}

    for _, target := range targets {
        go myPing(target)
    }
    time.Sleep(time.Second * 3) //Wait 3 seconds
}
```

If you run this program, it should output something like the following:

```
$ go run goroutine-icmp-probe.go
google.com OK, latency is 15.9587ms
cisco.com OK, latency is 163.6334ms
yahoo.com OK, latency is 136.3522ms
cern.ch OK, latency is 225.0571ms
```

To use a goroutine, you just need to add go in front of the function you want to call as a goroutine. The go statement indicates that the function can be executed in the background with its own stack and variables. The program then executes the line after the go statement and continues the flow normally without waiting for anything to return from the goroutine. As the ICMP probe request takes a few milliseconds to receive an ICMP response, the program will exit before it can print anything by the goroutines. Therefore, we need to add a sleep time of 3 seconds before finishing the program to make sure all the goroutines that send ICMP requests have received and printed the results. Otherwise, you won't be able to see any output, because the program will end before the goroutines finish printing the results.

If you want to wait until the goroutines end, Go has mechanisms to communicate and wait until they end. One simple one is using sync.WaitGroup. Let's now rewrite our previous example, removing the sleep time and adding `WaitGroup` to wait for the goroutines to finish. The following is the same example that waits until all goroutines end:

```go
import (
    "fmt"
    "sync"
    "github.com/go-ping/ping"
)


func myping(host string, wg *sync.WaitGroup) {
    defer wg.Done()
    p, err := ping.NewPinger(host)
    if err != nil {
        panic(err)
    }
    p.Count = 1
    p.SetPrivileged(true)
    if err = p.Run(); err != nil {
        panic(err)
    }
    stats := p.Statistics()
    fmt.Println(host, "OK, latency is", stats.AvgRtt)
}


func main() {
    var targets = []string{"yahoo.com", "google.com", "cisco.
com", "cern.ch"}

    var wg sync.WaitGroup
    wg.Add(len(targets))
    for _, target := range targets {
        go myping(target, &wg)
    }
    wg.Wait()
}
```

If you run the preceding code, it should end faster than the previous one because it does not sleep for 3 seconds; it only waits until all goroutines end, which should be less than half a second.

To allow `sync.WaitGroup` to work, you have to set a value to it at the beginning using `Add()`. In the preceding example, it adds 4, which is the number of goroutines that will run. Then, you pass the pointer of the variable to each goroutine function (`&wg`), which will be marked as `Done()` as the function ends using `defer` (explained in *Chapter 7*).

In the preceding example, we did not generate any communication between the goroutines, as they use the terminal to print. We only passed a pointer to the workgroup variable, called wg. If you want to communicate between goroutines, you can do that by using `channel`, which can be unidirectional or bidirectional.

More on goroutines can be found at the following links:

- *Google I/O 2012 - Go Concurrency Patterns*: `https://www.youtube.com/watch?v=f6kdp27TYZs`

- *Google I/O 2013 – Advanced Go Concurrency Patterns*: `https://www.youtube.com/watch?v=QDDwwePbDtw`

- More documentation on Goroutines can be found at `go.dev/doc/effective_go#concurrency`

Before going to the next section, let's summarize how we scale up in Python and Go. In Python, to make the right choice, use *Figure 8.1*:

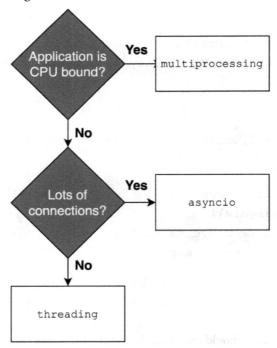

Figure 8.1 – Python decision-making for scaling your code

The diagram in *Figure 8.1* shows which Python library to use when scaling your code. If CPU bound, use multiprocessing. If you have too many connections with slow I/O, use asyncio, and if the number of connections is small, use threading.

For Go, there is only one option, which is a goroutine. Easy answer!

Let's now check how we can scale the system using schedulers and dispatchers.

Adding schedulers and job dispatchers

A scheduler is a system that selects a job for running, where a **job** can be understood as a program or part of code that requires running. A dispatcher, on the other hand, is the system that takes the job and places it in the execution queue of a machine. They are complementary, and in some cases, they are treated as the same system. So, for the purpose of this section, we are going to talk about some systems that can do both scheduling and dispatching jobs.

The main objective of using systems that can schedule and dispatch jobs is to gain scale by adding machines that can run more jobs in parallel. It is kind of similar to a single program using multiprocessing but with the difference that the new processes are being executed on another machine.

You could do lots of work to improve the performance of your program, but in the end, you will be bound by the machine's limitations, and if your application is CPU bound, it will be limited by the number of cores, the number of concurrent threads, and the speed of the CPU used. You could work hard to improve the performance of your code, but to grow more, the only solution is to add more CPU hardware, which can be accomplished by adding machines. The group of machines that are dedicated to running jobs for schedulers and dispatchers is normally called a **cluster**.

A cluster of machines that are ready to run jobs in parallel can be installed locally or can be installed in separate locations. The distance between the machines in a cluster adds latency to the communication between the machines and delays data synchronization. Quick synchronization between machines may or may not be relevant, depending on how fast the results are required and how they need to be combined if they depend on each other in time. Quicker results might require a local cluster. A more relaxed time frame for getting results would allow clusters to be located further apart.

Let's now discuss how we can use a classical scheduler and dispatcher.

Using classical schedulers and dispatchers

The classical scheduler and dispatcher would be any system that takes one job, deploys it to a machine in the cluster, and executes it. In a classical case, the job is just a program that is ready to run on a machine. The program can be written in any language; however, there are differences in how the installation would be compared between Python and Go.

Let's investigate what the differences are between using a cluster ready to run Python scripts and ready to run Go-compiled code.

Considerations for using Go and Python

If the program was written in Python, it is required that all machines in the cluster have the Python interpreter version that is compatible with the Python code. For instance, if code was written for Python 3.10, the CPython interpreter to be installed must be at least version 3.10. Another important point here is that all third-party libraries used in the Python script will also have to be installed on all machines. The version of each third-party library must be compatible with the Python script, as newer versions of a particular third-party library might break the execution of your code. You might need to maintain a table somewhere containing the versions of each third-party library used to avoid wrong machine updates. In conclusion, using Python complicates your cluster installation, management, and updates a lot.

On the other hand, using the Go language is much simpler for deploying in a cluster. You just need to compile the Go program to the same CPU architecture that the code will run. All third-party libraries used will be added to the compiled code. The versions of each third-party library used in your program will be controlled automatically by your local development environment with the go.sum and go.mod files. In summary, you don't need to install an interpreter on the machines and don't need to worry about installing or updating any third-party libraries, which is much simpler.

Let's now see a few examples of a scheduler and dispatcher for a cluster of machines.

Using Nomad

Nomad is an implementation for job scheduling that was built using Go and is supported by a company called HashiCorp (https://www.nomadproject.io/). Nomad is also very popular for scheduling and launching Docker containers in a cluster of machines, as we are going to see in the next section.

With Nomad, you are able to define a job by writing a configuration file that describes the job and how you want to run it. The job description can be written in any formatted file, such as YAML or TOML, but the default supported format is **HCL**. Once you have the job description completed, it is then translated to JSON, which will be used on the Nomad API (more details can be found at https://developer.hashicorp.com/nomad/docs/job-specification).

Nomad supports several task drivers, which allows you to schedule different kinds of programs. If you are using a Go-compiled program, you will have to use the Fork/Exec driver (further details are available at https://developer.hashicorp.com/nomad/docs/drivers/exec). Using the Fork/Exec driver, you can execute any program, including Python scripts, but with the caveat of having all third-party libraries and the Python interpreter previously installed on all machines of the cluster, which is not managed by Nomad and must be done on your own separately.

The following is an example of a job specification for an ICMP probe program:

```
job "probe-icmp" {
  region = "us"
  datacenters = ["us-1", "us-12"]
```

```
type = "service"
update {
  stagger     = "60s"
  max_parallel = 4
}
task "probe" {
  driver = "exec"
  config {
    command = "/usr/local/bin/icmp-probe"
  }
  env {
    TARGETS = "cisco.com,yahoo.com,google.com"
  }
  resources {
    cpu    = 700 # approximated in MHz
    memory = 16 # in MBytes
  }
}
}
```

Note that the preceding program example is called `icmp-probe` and would have to accept the operating system environment variable as input. In our example, the variable is called `TARGETS`.

Once you have defined your job, you can dispatch it by issuing the `nomad job dispatch <job-description-file>` command (more details can be found at `developer.hashicorp.com/nomad/docs/commands/job/dispatch`).

Let's now check how we could use another popular scheduler.

Using Cronsun

Cronsun is another scheduler and dispatcher that works in a similar way to the popular Unix cron but for multiple machines. The goal of Cronsun is to be easy and simple for managing jobs on lots of machines. It is developed in the Go language but can also launch jobs in any language by invoking a shell on the remote machine, such as in the Nomad `Fork/Exec` driver (more details can be found at `https://github.com/shunfei/cronsun`). It also has a graphical interface that allows easy visualization of the running jobs. Cronsun was built and designed based on another Go third-party package, called `robfig/cron` (`https://github.com/robfig/cron`).

Using Cronsun, you will be able to launch several jobs on multiple machines, but there is no machine cluster management like in Nomad. Another important point is Cronsun does not work with Linux containers, so it purely focuses on executing Unix shell programs in the remote machine by doing process forking.

Let's now look at a more complex scheduler.

Using DolphinScheduler

DolphinScheduler is a complete system for scheduling and dispatching jobs that is supported by the Apache Software Foundation. It has many more features compared to Nomad and Cronsun, with workflow capabilities that allow a job to wait for input from another job before executing. It also has a graphical interface that helps to visualize running jobs and dependencies (more details can be found at `https://dolphinscheduler.apache.org/`).

Although DolphinScheduler is primarily written in Java, it can dispatch jobs in Python and Go. It is much more complex and has many capabilities that might not be necessary for your requirements to scale up.

There are several other job schedulers and dispatchers that you could use, but some of them are used for specific languages, such as Quartz.NET, used for .NET applications (`https://www.quartz-scheduler.net/`), and Bree, used for Node.js applications (`https://github.com/breejs/bree`).

Let's see now how we can use big data schedulers and dispatchers for carrying out computation at scale in network automation.

Working with big data

There are specific applications that require lots of CPU for data processing. These applications require a system that allows running code with very specialized algorithms focused on data analysis. These are normally referred to as systems and applications for **big data**.

Big data is a collection of datasets that are too large to be analyzed on just one computer. It is a field that is dominated by data scientists, data engineers, and artificial intelligence engineers. The reason is that they normally analyze a lot of data to extract information, and their work requires a system that scales up a lot in terms of CPU processing. Such scale can only be achieved by using systems that can schedule and dispatch jobs over many computers in a cluster.

The algorithm model used for big data is called **MapReduce**. A MapReduce programming model is used to implement analysis on large datasets using an algorithm that runs on several machines in a cluster. Originally, the term MapReduce was related to a Google product, but now it is a term used for programs that deal with big data.

The original paper published by Jeffrey Dean and Sanjay Ghemawat called *MapReduce: Simplified Data Processing on Large Clusters* is a good reference and good reading to dive deeper into the subject. The paper is public and can be downloaded from the Google Research page at `https://static.googleusercontent.com/media/research.google.com/en//archive/mapreduce-osdi04.pdf`.

Let's see how we can use big data in our network automation.

Big data and network automation

Big data is used in network automation to help with traffic engineering and optimization. MapReduce is used to calculate better traffic paths over a combination of traffic demands and routing paths. Traffic demands are collected and stored using the IP source and IP destination, then MapReduce is used to calculate a traffic demand matrix. For this work, routing and traffic information is collected from all network devices using BGP, SNMP, and a flow-based collection such as `sflow`, `ipfix`, or `netflow`. The data collected is normally big and real-time results are required to allow for proper network optimization and traffic engineering on time.

One example would be the collection of IP data flow from the transit routers and peering routers (discussed in *Chapter 1*). This flow information would then be analyzed in conjunction with the routing information obtained from the routers. Then, a better routing policy can be applied in real time to select better external paths or network interfaces that are less congested.

Let's now investigate some popular systems that can be used for big data.

Using systems for big data

The two most popular open source systems for big data are **Apache Hadoop** (`https://hadoop.apache.org/`) and **Apache Spark** (`https://spark.apache.org/`). Both systems are supported and maintained by the Apache Software Foundation (`https://www.apache.org/`) and are used to build large cluster systems to run big data.

The difference between Hadoop and Spark is related to how they perform big data analysis. Hadoop is used for batch job scheduling without real-time requirements. It uses more disk capacity and the response time is more relaxed, so the cluster machines don't need to be local, and the machines need to have large disks. On the other hand, Spark uses more memory and less disk space, the machines need to be located closer, and the response time is more predictable, therefore it is used for real-time applications.

For our network automation on traffic analysis, either option can be used, but Spark would be preferred for faster and more periodic results. Hadoop would be used to generate monthly and daily reports, but not to interact with real-time routing policies.

Let's now look at a common problem with having your own cluster.

Resource allocation and cloud services

One of the problems with using Hadoop and Spark is that you will need to create your own cluster of machines. That means installing and maintaining the hardware and operating system software. But that is not the main problem. The main problem is that resource utilization will vary throughout the day and the year.

As an example, imagine you are using a big data system in your company to calculate the best path for a particular group of routers during the day. The problem is the collected data to be analyzed will change; in busy hours, you will need more CPU processing to calculate compared to non-busy hours. The difference can be hundreds of CPUs, which will lead to lots of idle CPU hours at the end of the month.

How do you solve this issue? By using a cloud-based service provider to allocate machines for your cluster. With it, you can add and remove machines during the day and throughout the week, allowing growth when needed and releasing computing power when not needed. One example is to use AWS' product called **Elastic MapReduce** (**EMR**), which can be used with easy machine allocation for your cluster, scaling up and down by software (more details can be found at `https://aws.amazon.com/emr/`). Similar services can be obtained from other cloud service providers, such as Google, Oracle, or Microsoft.

One important point to observe is that big data systems do not allow running any program or language, but only code that has the MapReduce concept capabilities. So, it is much more specific compared to Nomad or Cronsun, and focuses only on data analysis.

Let's now check how we can scale using microservices and Linux containers.

Using microservices and containers

When software is built based on a combination of small, independent services, we normally say the software was built using microservices architecture. Microservices architecture is a way to develop applications by combining small services that might belong or not to the same software development team.

The success of this approach is due to the isolation between each service, which is accomplished by using Linux containers (described in *Chapter 2*). Using Linux containers is a good way to isolate memory, CPU, networks, and disks. Each Linux container can't interact with other Linux containers in the same host unless a pre-defined communication channel is established. The communication channels of a service have to use well-documented APIs.

The machine that runs microservices is normally called a **container host** or just a host. A host can have multiple microservices that may or may not communicate with each other. A combination of hosts is called a cluster of container hosts. Some orchestration software is able to spawn several copies of a service in one host or different hosts. Using microservices architecture is a good way to scale your system.

One very popular place to build and publish a microservice is **Docker** (https://www.docker.com/). A Docker container is normally referred to as a service that is built using a Linux container. A Docker host is where a Docker container can run, and in a similar way, a Docker cluster is a group of hosts that can run Docker containers.

Let's see now how we can use Docker containers to scale our code.

Building a scalable solution by example

Let's build a solution using microservices architecture by creating our own Docker container and then launching it multiple times. Our service has a few requirements, as follows:

- It needs to have an API to accept requests
- The API needs to accept a list of targets
- An ICMP probe will be sent to each target to verify latency concurrently
- The API will respond using HTTP plain text
- Each service can accept up to 1,000 targets
- The timeout for each ICMP probe must be 2 seconds

Based on these requirements, let's write some code that will be used in our service.

Writing the service code

With the previous requirements, let's write some code in Go to build our service. We are going to use the Go third-party package for ICMP that we used before in this chapter called go-ping/ping, and sync.WaitGroup to wait for the goroutines to end.

Let's break the code into two blocks. The second block of code is as follows, describing the probeTargets() and main() functions:

```go
func probeTargets(w http.ResponseWriter, r *http.Request) {
    httpTargets := r.URL.Query().Get("targets")
    targets := strings.Split(httpTargets, ",")
    if len(httpTargets) == 0 || len(targets) > 1000{
        fmt.Fprintf(w, "error: 0 < targets < 1000\n")
        return
    }

    var wg sync.WaitGroup
    wg.Add(len(targets))
```

```
    for _, target := range targets {
        log.Println("requested ICMP probe for", target)
        go probe(target, w, &wg)
    }
    wg.Wait()
}

func main() {
    http.HandleFunc("/latency", probeTargets)
    log.Fatal(http.ListenAndServe(":9900", nil))
}
```

The preceding block represents the last two functions of our service. In the main() function, we just need to call http.HandleFunc, passing the API reference used for the GET method and the name of the function that will be invoked. Then, http.ListenAndServe is called using port 9900 to listen for API requests. Note that log.Fatal is used with ListenAndServe because it should never end unless it has a problem. The following is an API GET client request example:

```
GET /latency?targets=google.com,cisco.com HTTP/1.0
```

The preceding API request will call probeTargets(), which will run the loop invoking the goroutines (called probe()) two times, which will send ICMP requests to google.com and cisco.com.

Let's now have a look at the last block of code containing the probe() function:

```
func probe(host string, w http.ResponseWriter, wg *sync.
WaitGroup) {
    defer wg.Done()
    p, err := ping.NewPinger(host)
    if err != nil {
        fmt.Fprintf(w, "error ping creation: %v\n", err)
        return
    }

    p.Count = 1
    p.Timeout = time.Second * 2
    p.SetPrivileged(true)
    if err = p.Run(); err != nil {
        fmt.Fprintf(w, "error ping sent: %v\n", err)
```

```
        return
    }
    stats := p.Statistics()
    if stats.PacketLoss == 0 {
        fmt.Fprintf(w, "%s latency is %s\n", host, stats.
AvgRtt)
    } else {
        fmt.Fprintf(w, "%s no response timeout\n", host)
    }
}
```

Note that the `probe()` function does not return a value, a `log` message, or a print message. All messages, including errors, are returned to the HTTP client requesting the ICMP probes. To allow the return to the client, we have to use the `fmt.Fprintf()` function, passing the reference w, which points to an `http.ResponseWriter` type.

Before we continue with our example, let's make a modification to our `main()` function to allow reading the port number from the operating system environment variable. So, the service can be called with different port numbers when being invoked, just needing to change the operating system environment variable called PORT, as shown here:

```
func main() {
    listen := ":9900"
    if port, ok := os.LookupEnv("PORT"); ok {
        listen = ":" + port
    }

    http.HandleFunc("/latency", probeTargets)
    log.Fatal(http.ListenAndServe(listen, nil))
}
```

Let's now build our Docker container using a Dockerfile.

Building our Docker container

To build the Docker container, we are going to use Dockerfile definitions. Then, we just need to run `docker build` to create our container. Before you install the Docker engine in your environment, check the documentation on how to install it at `https://docs.docker.com/engine/install/`.

The following is the Dockerfile used in our example of an ICMP probe service:

```
FROM golang:1.19-alpine
WORKDIR /usr/src/app
COPY go.mod go.sum ./
RUN go mod download && go mod verify
COPY icmp-probe-service.go ./
RUN go build -v -o /usr/local/bin/probe-service
CMD ["/usr/local/bin/probe-service"]
```

To build the Docker container, you just need to run `docker build . -t probe-service`. After running the build, you should be able to see the image by using the `docker image` command, as follows:

```
% docker images
REPOSITORY         TAG     IMAGE ID  CREATED           SIZE
probe-service      latest  e9c2      About a minute ago  438MB
```

The Docker container name is `probe-service` and you can run the service by using the following command:

```
docker run -p 9900:9900 probe-service
```

To listen to a different port, you need to set the `PORT` environment variable. An example for port `7700` is as follows:

```
docker run -e PORT=7700 -p 7700:7700 probe-service
```

Note that you could map different host ports to port `9900` if you want to run multiple services in the same host without changing the port that the container listens to. You just need to specify a different port for the host when mapping, as in the following example running three services on the same machine:

```
% docker run -d -p 9001:9900 probe-service
% docker run -d -p 9002:9900 probe-service
% docker run -d -p 9003:9900 probe-service
```

Running the preceding three commands will start three services on the host ports: `9001`, `9002`, and `9003`. The service inside the container still uses port `9900`. To check the services running in a host, use the `docker ps` command, as follows:

```
% docker ps
CONTAINER ID    IMAGE           COMMAND
```

```
        CREATED          STATUS            PORTS
          NAMES
6266c895f11a     probe-service     "/usr/local/bin/prob..."
2 minutes ago    Up 2 minutes      0.0.0.0:9003->9900/tcp
gallant_heisenberg
270d73163d19     probe-service     "/usr/local/bin/prob..."
2 minutes ago    Up 2 minutes      0.0.0.0:9002->9900/tcp
intelligent_clarke
4acc6162e821     probe-service     "/usr/local/bin/prob..."
2 minutes ago    Up 2 minutes      0.0.0.0:9001->9900/tcp
hardcore_bhabha
```

The preceding output shows that there are three services running on the host, listening to ports 9001, 9002, and 9003. You can access the APIs for each of them and probe up to 3,000 targets, 1,000 per service.

Let's now see how we can automate launching multiple services using Docker Compose.

Scaling up using Docker Compose

Using Docker Compose will help you to add services that will run at the same time without needing to invoke the docker run command. In our example, we are going to use Docker Compose to launch five ICMP probe services. The following is the Docker Compose file example in YAML format (described in *Chapter 4,*):

```yaml
version: "1.0"
services:
  probe1:
    image: "probe-service:latest"
    ports: ["9001:9900"]
  probe2:
    image: "probe-service:latest"
    ports: ["9002:9900"]
  probe3:
    image: "probe-service:latest"
    ports: ["9003:9900"]
  probe4:
    image: "probe-service:latest"
    ports: ["9004:9900"]
  probe5:
```

```
        image: "probe-service:latest"
        ports: ["9005:9900"]
```

To run the services, just type docker compose up -d, and to stop them, just run docker compose down. The following is an example of the output of the command:

```
% docker compose up -d
[+] Running 6/6
⋮⋮ Network probe-service_default    Created   1.4s
⋮⋮ Container probe-service-probe5-1  Started   1.1s
⋮⋮ Container probe-service-probe4-1  Started   1.3s
⋮⋮ Container probe-service-probe3-1  Started   1.5s
⋮⋮ Container probe-service-probe1-1  Started   1.1s
⋮⋮ Container probe-service-probe2-1  Started
```

Now, let's see how we can scale up using multiple machines with a Docker container.

Scaling up with clusters

To scale even more, you could set up a cluster of Docker host containers. This will allow you to launch thousands of services, allowing our ICMP probe service to scale to millions of targets. You could build the cluster yourself by managing a group of machines and running the services, or you could use a system to do all that for you.

Let's now investigate a few systems that are used to manage and launch services for a cluster of machines running container services.

Using Docker Swarm

With **Docker Swarm**, you are able to launch containers on several machines. It is easy to use because it only requires installing Docker. Once you have installed it, it is easy to create a Docker Swarm cluster. You just have to run the following command:

```
Host-1$ docker swarm init
Swarm initialized: current node (9f2777swvj1gmqegbxabahxm3) is
now a manager.
To add a worker to this swarm, run the following command:
    docker swarm join --token SWMTKN-1-1gdb6i88ubq5drnigbw
q2rh51fmyordkkpljjtwefwo2nk3ddx-6nwz531o61qtkun4gagvrl7ws
192.168.86.158:2377
```

```
To add a manager to this swarm, run 'docker swarm join-token
manager' and follow the instructions.
```

Once you have started the first Docker Swarm host, it will then take the lead place, and to add another host, you just need to use the `docker swarm join` command. To avoid any host joining the Docker Swarm cluster, a token is used. The preceding example starts with `SWMTKN-1`. Note that a host in a Docker Swarm cluster is also called a **node**. So, let's add more nodes to our cluster:

```
host-2$ docker swarm join --token SWMTKN-1-1gdb6i88ubq5drnig
bwq2rh51fmyordkkpljjtwefwo2nk3ddx-6nwz531o61qtkun4gagvrl7ws
192.168.86.158:2377

host-3$ docker swarm join --token SWMTKN-1-1gdb6i88ubq5drnig
bwq2rh51fmyordkkpljjtwefwo2nk3ddx-6nwz531o61qtkun4gagvrl7ws
192.168.86.158:2377

host-4$ docker swarm join --token SWMTKN-1-1gdb6i88ubq5drnig
bwq2rh51fmyordkkpljjtwefwo2nk3ddx-6nwz531o61qtkun4gagvrl7ws
192.168.86.158:2377
```

Now, we have four nodes in the cluster, with `host-1` as the leader. You can check the status of the cluster nodes by typing the following command:

```
$ docker node ls
ID              HOSTNAME    STATUS    AVAILABILITY MANAGER
9f2777swvj*     host-1      Ready     Active       Leader
a34f25affg*     host-2      Ready     Active
7fdd77wvgf*     host-4      Ready     Active
8ad531vabj*     host-3      Ready     Active
```

Once you have your cluster, you can launch a service by running the following command:

```
$ docker service create --replicas 1 --name probe probe-service

7sv66ytzq0te92dkndz5pg5q2
overall progress: 1 out of 1 tasks
1/1: running
[==================================================>]
verify: Service converged
```

In the preceding example, we just launched a Swarm service called `probe` using the `probe-service` image, the same image used in previous examples. Note that we've only launched one replica to later show how easy it is to scale up. Let's check now how the service is installed by running the following command:

```
$ docker service ls
ID             NAME   MODE         REPLICAS  IMAGE
7sv66ytzq0     probe  replicated   1/1       probe-service:latest
```

Let's now scale up for 10 probes by running the following command:

```
$ docker service scale probe=10
probe scaled to 10
overall progress: 10 out of 10 tasks
1/10: running
[==================================================>]
2/10: running
[==================================================>]
3/10: running
[==================================================>]
4/10: running
[==================================================>]
5/10: running
[==================================================>]
6/10: running
[==================================================>]
7/10: running
[==================================================>]
8/10: running
[==================================================>]
9/10: running
[==================================================>]
10/10: running
[==================================================>]
verify: Service converged
```

Now, if you check the service, it will show 10 replicas, as in the following command:

```
$ docker service ls
ID            NAME    MODE         REPLICAS IMAGE
7sv66ytzq0    probe   replicated   10/10     probe-service:latest
```

You can also check where each replica is running by running the following command:

```
$ docker service ps probe
ID       NAME       IMAGE                   NODE     DESIRED STATE
y38830   probe.1    probe-service:latest    host-1   Running Running
v4493s   probe.2    probe-service:latest    host-2   Running Running
zhzbnj   probe.3    probe-service:latest    host-3   Running Running
i84s4g   probe.4    probe-service:latest    host-3   Running Running
3emx3f   probe.5    probe-service:latest    host-1   Running Running
rd1vpl   probe.6    probe-service:latest    host-2   Running Running
ploq0w   probe.7    probe-service:latest    host-3   Running Running
ro0foo   probe.8    probe-service:latest    host-4   Running Running
16prr4   probe.9    probe-service:latest    host-4   Running Running
dwdr43   probe.10   probe-service:latest    host-1   Running Running
```

As you can see in the output of the preceding command, there are 10 probes running as replicas on nodes host-1, host-2, host-3, and host-4. You can also specify where you want the replica to run, among other parameters. In this example, we are able to scale up our ICMP probe service to 10,000 targets by using 4 hosts.

One important point we missed on these commands was allocating the ports to listen for our replicas. As replicas can run in the same host, they can't use the same port. We then need to make sure each replica is assigned with a different port number to listen to. A client accessing our probe-service cluster needs to know the IP addresses of the hosts and the port numbers that are listening before connecting for requests.

A better and more controlled way to deploy Docker Swarm is to use a YAML configuration file like we did when using Docker Compose. More details on the configuration file for Docker Swarm can be found at https://github.com/docker/labs/blob/master/beginner/chapters/votingapp.md.

More documentation on Docker Swarm can be found at https://docs.docker.com/engine/swarm/swarm-mode/.

Now let's investigate how to use multiple hosts using Kubernetes.

Using Kubernetes

Kubernetes is perhaps one of the most popular systems to manage the microservices architecture in a cluster. Its popularity is also due to it being backed by the **Cloud Native Computing Foundation** (`https://www.cncf.io/`), which is part of the **Linux Foundation** (`https://www.linuxfoundation.org/`). Large companies use Kubernetes, such as Amazon, Google, Apple, Cisco, and Huawei, among others.

Kubernetes provides many more capabilities than Docker Swarm, such as service orchestration, load-balancing, service monitoring, self-healing, and auto-scaling by traffic, among other features. Despite the large community and vast capabilities, you might not want to use Kubernetes if your requirement is simple and needs to scale in large quantities. Kubernetes provides a lot of capabilities that might be an overhead to your development. For our `probe-service`, I would not recommend using Kubernetes, because it is too complex for our purposes of ICMP probing targets.

You can also use a Docker Compose file to configure Kubernetes, which is done by using a service translator such as Kompose (`https://kompose.io/`). More details can be found at `https://kubernetes.io/docs/tasks/configure-pod-container/translate-compose-kubernetes/`.

If you want to start using Kubernetes, you can find plenty of examples and documentation on the internet. The best place to start is at `https://kubernetes.io/docs/home/`.

Let's now check how we can use another cluster based on Nomad.

Using Nomad

Nomad is also used to implement Docker clustering (`https://www.nomadproject.io/`). Nomad also has several capabilities that are comparable to Kubernetes, such as monitoring, self-healing, and auto-scaling. However, the features list is not as long and complete as Kubernetes.

So, why would we use Nomad instead of Kubernetes? There are three main reasons that you might want to use Nomad, as listed here:

- Simpler to deploy and easy to configure in comparison to Kubernetes.
- Kubernetes can scale up to 5,000 nodes with 300,000 containers. Nomad, on the other hand, is able to scale to 10,000 nodes and more than 2 million containers (`https://www.javelynn.com/cloud/the-two-million-container-challenge/`).
- Can support other services besides Linux containers, such as **QEMU** virtual machines, Java, Unix processes, and Windows containers.

More documentation on Nomad can be found at `https://developer.hashicorp.com/nomad/docs`.

Let's now have a brief look at how to use microservice architectures provided by cloud service providers.

Using cloud service providers

There are also proprietary solutions that are provided by cloud service providers, such as **Azure Container Instances**, **Google Kubernetes Engine** (GKE), and Amazon **Elastic Container Service** (**ECS**). The advantage of using a cloud service provider is you don't need physical machines in your infrastructure to create the cluster. There are also products where you don't even need to care about the cluster and the nodes in it, such as a product from Amazon called AWS Fargate. With AWS Fargate, you just need the Docker container published in a Docker registry and a service specification without the need to specify the nodes or the cluster.

I hope this section has given you a good idea of how to scale your code by using Linux containers and host clustering. Microservice architecture is a hot topic that has gotten lots of attention from developers and cloud service providers in recent years. Several acronyms might be used to describe this technology, but we have covered the basics here. You now have enough knowledge to dive even deeper into this subject.

Summary

This chapter has shown you a good summary of how you can improve and use systems to scale your code. We also demonstrated how we can use standard and third-party libraries to add capabilities to our code to scale.

Now, you are probably much more familiar with the technologies that you could use to interact with large networks. You are now in a better position to choose a language, a library, and a system that will support your network automation to scale to handle thousands or even millions of network devices.

In the next chapter, we are going to cover how to test your code and your system, which will allow you to build solutions for network automation that are less prone to failures.

Part 3:
Testing, Hands-On, and Going Forward

The third part of the book will discuss what has to be considered when building a framework for testing your code and how to do so, We will do some real hands-on testing and, finally, describe what to do to move forward in the network automation realm. We will provide the details on creating a testing framework and do hands-on work using an emulated network, which will help to put into practice all the information learned in previous parts.

This part has the following chapters:

- *Chapter 9, Network Code Testing Frameworks*
- *Chapter 10, Hands-On and Going Forward*

9

Network Code Testing Framework

One important aspect when developing code is to add testing; we discussed and reviewed some code test strategies in *Chapter 5, Do's and Don'ts for Network Programming*. But we have not investigated techniques that are unique to network automation, such as building a network testing environment where we can do some real testing with our network automation code.

This chapter will focus on techniques for building a network testing framework that can be used for testing your network automation code. We are also going to look into advanced techniques that can be added to make your testing framework even more useful and reliable.

Here are the topics we are going to cover in this chapter:

- Using software for testing
- Using device emulation
- Connecting devices for testing
- Using advanced testing techniques

By the end of this chapter, you should have enough information to build and use a testing framework that will add significant value to your network automation project.

Technical requirements

The source code described in this chapter is stored in this book's GitHub repository at https://github.com/PacktPublishing/Network-Programming-and-Automation-Essentials/tree/main/Chapter09.

Using software for testing

Some companies, when buying network devices, buy additional devices to be used for testing purposes. These extra devices are normally installed in a separate environment to replicate part of the production network for testing. Before one test is performed, the devices are connected and configured to replicate a particular part of the network. Once the tests are finished, the setup is then removed and another configuration with different connections may take place to perform tests for another part of the network. There are several reasons why these tests are necessary, such as testing new software, validating a new configuration, verifying an update, checking performance, qualifying a new network design, and testing new functionality, among others.

But the main problem is that the testing environment is costly, slow to set up, and cannot be used by multiple test engineers in parallel. It also demands the physical presence of a specialized technician who will eventually have to sort new cable connections, perform hardware updates, add or remove network cards, and sometimes update the operating systems of the devices.

The physical testing environment is ultimately inevitable, but a few tests can be performed by using software instead of physical hardware. The tests that can be performed by software will depend on the requirement of the test. Tests that evaluate software configuration, verify design concepts, validate routing behavior, and validate new features and perhaps router stability might be performed by software.

In addition, software can be used to make connections between network devices, which also speeds up the setup process. But one area won't be able to be tested, which is network stress and performance, such as measuring maximum throughput or capacity.

There are several techniques we can use to use software for network testing, and most of them will be done using simulation and emulation. But what are the differences between emulation and simulation? Let's discuss this now.

Differences between emulation and simulation

Emulation and simulation have their meanings commonly mixed it up. Although mistaking them is not that important, it is good to understand their meaning so that you can understand their limitations and capabilities when using them.

In the case of emulation, it is when you use software to mimic the physical aspects of the entity you want to test. So, in our network automation, it would be a router, a switch, or a network connection (link).

Therefore, using a router emulation implies that all necessary hardware, such as network ports, console ports, CPU, and memory, must be mocked by software so that the operating system of the router can run seamlessly as if it were running in real hardware. One example of a router emulator is **Dynamips** (there are more details at `https://github.com/GNS3/dynamips`).

On the other hand, a simulator is built to mimic some functionalities of the entity that you want to test. In the case of a router, normally, only particular functions are simulated, not all the functions of the router. Because of the characteristics of being smaller, a simulator can accomplish results much faster and can scale up to thousands in comparison to an emulator. Two popular examples of software used for simulating networks are **ns-3** (`https://www.nsnam.org/`) and **NetworkX** (`https://networkx.org/`).

Now that we know the differences between emulation and simulation, let's dive a bit deeper into emulation.

Using device emulation

The best use case for using emulation in our network automation would be for routers. With router emulation, we can test several features of the router without having it physically. But router emulation is perhaps the hardest to accomplish and the costliest in terms of resources. As an example, let's explore how a popular Cisco router emulator works, called Dynamips. *Figure 9.1* represents a Cisco router being emulated by using Dynamips on a Linux host:

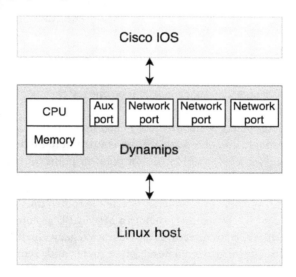

Figure 9.1 – Cisco router emulation

As illustrated, Dynamips is a software layer that emulates the hardware for Cisco routers. Dynamips can emulate some Cisco hardware such as network ports, CPU, memory, auxiliary ports, and console ports. Dynamips was created by Christophe Fillot in 2005 to emulate the MIPS processor architecture for Cisco routers. Today, Dynamips is supported and maintained by the GNS network simulation team, more details of which can be found at `https://github.com/GNS3/dynamips`.

Dynamips works like a virtual machine in that it will only run a Cisco operating system. To emulate a MIPs processor, Dynamips consumes a lot of CPU and memory. For instance, to run a legacy Cisco router 7200, Dynamips will allocate a minimum of 256 MB of RAM, plus 16 MB of cache.

CPU is also heavily used to run the router by translating instruction by instruction. Earlier versions of Dynamips overloaded the CPU host, but with the introduction of a capability called **idle-PC**, the CPU consumption was reduced by a significant amount.

Other routers can be emulated but will need an emulator that provides the necessary hardware emulation for the CPU platform you require. Juniper routers can be emulated by using Juniper Olive. Juniper Olive is FreeBSD modified to load the Juniper router operational system called JunOS. With emulation, you can also run legacy routers by using a hardware emulator that can provide legacy CPU architecture.

The following diagram illustrates a system running with four emulations, two Cisco routers, one Juniper router, and one legacy OpenWRT router:

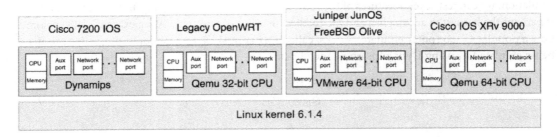

Figure 9.2 – Four emulations on one system

The connection between these routers in the preceding diagram is created on the operating system host. The host can provide a more complex software link emulation to provide connectivity or just copy traffic from one end and send it to another end for point-to-point connections. More on these connections will be explained later in this chapter in the *Connecting devices for testing* section.

Note that with emulation, it is possible to completely isolate the router, thus providing a completely different architecture. In *Figure 9.2*, Dynamips provides a MIPS CPU architecture to emulate Cisco 7200, Qemu provides a 32-bit CPU architecture to emulate a legacy OpenWRT router, VMware provides a 64-bit x86 CPU architecture to emulate a Juniper router, and Qemu provides a 64-bit x86 CPU architecture to emulate a Cisco XRv 9000.

The host that's used for emulating these routers is Linux with a 6.1.4 kernel, but it could be another kernel or another operating system such as Windows that's capable of running the emulators. The consumption of CPU and memory for *Figure 9.2* is quite high – Cisco 9000 requires at least 4 vCPUs and 16 GB of RAM, Juniper requires a minimum of 2 vCPUs and 512 MB of RAM, Legacy OpenWRT requires a minimum of 1 vCPU, and Cisco 7200 requires a minimum of 2 vCPUs and around 300 MB.

Therefore, creating a large network using router emulation is hard, and perhaps impossible because of limited resources. One way to scale up the emulation is to share the hardware drivers, memory, and CPU by using some sort of operating system isolation, such as using Linux containers or FreeBSD jails. But with container setup, you will have to use the same version of the kernel and the same CPU architecture for all routers. Therefore, if your router runs on an ARM processor, and your host is an x86 processor, Linux containers won't work. To work, the containers and the host must use the same CPU architecture.

Now, let's have a look at how to scale emulation using containers.

Scaling up emulation with containers

Emulation will scale up if you share resources from the host dynamically. However, that requires your router to run as a program on a Linux host that can be isolated as a container. That is perhaps a big limitation as most commercial routers do not run on Linux and can't be containerized. Choosing open source routers gives you the advantage of it being easy to move to a container-based architecture. Some big companies have chosen to move away from commercial operating system routers and migrate to a Linux-based routing architecture, which facilitates the creation of the emulated network.

Despite the commercial limitations, some vendors provide versions that can be containerized, such as Arista, Cisco, Juniper, and Nokia. This includes Cisco XRv and CSRv versions, Juniper vMX and vQFX, Arista vEOS, and Nokia VSR. One project that explores these capabilities is **vrnetlab** (there are more details at `github.com/vrnetlab/vrnetlab`).

There are a bunch of network tests that can be done, even if the routers are not the same as production, such as network design tests, topology migration tests, IP filter-based tests, and topology failover tests, among others. The reason is that most topologies run standard protocols and can be translated into an open source network platform. If you are using SDN and OpenFlow, that is also true.

The following diagram illustrates how you could run four emulated routers using containers:

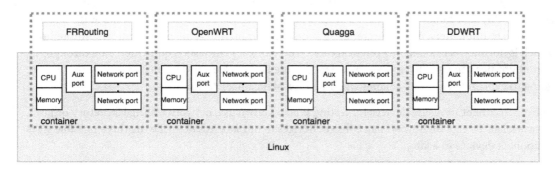

Figure 9.3 – Running emulation with containers

As you can see in the preceding figure, the containers share CPU, memory, and network ports with the Linux host (green rectangle), but with isolation within the container. Each container isolates the shared resources from other containers, but they use the same Linux kernel, same drivers, and same CPU architecture. Different runtime libraries can be used, but the kernel and CPU architecture must be the same. Each container will have its own routing table, and programs running on the same container will share the same routing table but will not share the routing tables between containers unless using a routing protocol.

You could also run a virtual machine inside a container, but then you are not saving resources, and the limitations shown previously are the same. So, if you want to scale up, you have to share the hardware resources with all containers, not emulate another layer like in *Figure 9.2*.

In the example in *Figure 9.3*, there are four routers – one FRRouting, one OpenWRT, one Quagga, and one DD-WRT. All these routers are open source and can be containerized. But they are not necessarily one program running, but instead a group of programs. Quagga and FFRouting run several programs that do different tasks, such as `bgpd`, `ospfd`, and `zebra`. References to these open source routers can be obtained from the following sources:

- FRRouting: `https://frrouting.org/`
- OpenWRT: `https://openwrt.org/docs/guide-user/virtualization/lxc`
- Quagga: `https://www.nongnu.org/quagga/`
- DD-WRT: `https://dd-wrt.com/`

You will need some connectivity capability to be able to connect the emulated routers. Now, let's discuss the techniques we can use to connect devices in a network.

Connecting devices for testing

Ensuring the connectivity of our devices for tests is important for obtaining a proper network environment for testing. There are several different ways to connect devices for testing, such as physical cables and software. Physical cables always have two caveats – they require technical personnel at the site and take time to implement. Via software, there is only one limitation – the maximum data throughput, which is normally a fraction of a physical cable. Therefore, if your tests require high data throughput, you might need to use physical wires instead. There is a workaround to this limitation that we are going to explain later in this chapter when we look at advanced techniques.

The environment for using devices for testing is also known as a network testing laboratory or just a network lab. To explain how we can connect devices in our lab, let's describe the three possible ways to connect devices in a lab.

Using physical wires to connect

In a test environment, a physical connection normally consists of wires that connect the ports of network devices. They are normally optical cables, coaxial cables, or twisted pair cables. If you have two devices, the cables are simple and will pass from one device to the other. However, if you are planning to have a lab with several racks and dozens of devices, you might want to use a patch cord and a patch panel instead of passing wires through the racks. The idea of using a patch panel is that the technician will only need to use patch cords to connect the devices, which makes the connection setup a bit faster and easier to remove later.

It is important to understand how the patch panel and patch cords work in the physical lab because it will help us to understand the software version later. The following diagram illustrates the patch panels for connecting two racks with four routers each:

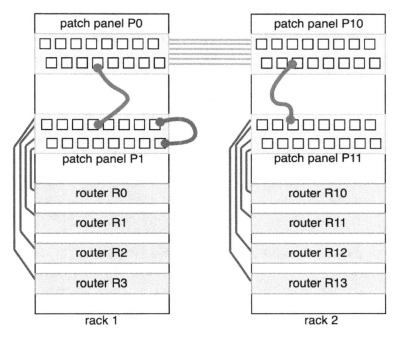

Figure 9.4 – Connecting routers using patch panels

Note that in the preceding figure, the orange and blue wires represent permanent wires and will never be removed. The red and green lines represent the patch cords that are used to connect devices but can be removed and reconnected easily for a different topology setup. The blue lines in rack 1 are connecting routers R0, R1, R2, and R3 to patch panel P1, similar to rack 2, which connects routers R10, R11, R12, and R13 to patch panel P11. The orange lines represent the permanent wires that connect patch panel P0 to patch panel P10.

Whenever a topology setup is required, the technician just needs patch cords to set up connections between the routers. The number of ports on each patch panel will depend on the number of available network ports on each router. As an example, for *Figure 9.4*, let's suppose that each router in rack 1 has five network ports available. Therefore, patch panel P1 needs to have at least 20 ports to allow connections to all routers on rack 1.

In *Figure 9.4*, there are three patch cords. The green one in rack 1 connects two devices inside rack 1, which could be, for instance, R0 and R1. The other two red patch cords are used to interconnect devices between rack 1 and rack 2, which could be, for instance, between R0 and R10.

Now, let's see how we can link devices using software connections.

Using software to connect

For the explanation in this subsection, we are going to assume all routers are software-emulated routers. A hybrid setup with software and real devices will be explained in the next subsection.

Several software techniques can be used to interconnect emulated routers, and they will also depend on the operating system that is used as the host. For our example, we are going to use Linux as the host. For Windows, FreeBSD, or macOS, you might need different techniques.

The methods to connect emulated routers will also depend on which emulation you are using. They may differ depending on whether you are using Dynamips, VirtualBox, VMware, Qemu, or Linux containers.

Let's explore a few methods to connect emulated routers using Linux.

Using TUN/TAP interfaces

In Linux, TUN/TAP interfaces are software interfaces that are used to receive and send network traffic, but they are not connected to any network. The interface is called TUN/TAP because the device can be either configured to work only on layer 3, which is called TUN mode, or on layer 2, which is called TAP interface mode. Both modes use the same Linux device driver (accessible via /dev/net/ tun), just with a different flag. The flag to use TAP mode is IFF_TAP, whereas the flag to use TUN is IFF_TUN. More details on the kernel driver for TUN/TAPc can be found at https://www. kernel.org/doc/html/v5.8/networking/tuntap.html.

Linux provides an easy interface for creating and removing TUN/TAP interfaces; you can use the ip tuntap command for this. The following is an example of creating a tap interface:

```
claus@dev:~$ sudo ip tuntap add dev tap0 mode tap
claus@dev:~$ sudo ip link set tap0 up
claus@dev:~$ ip link show tap0
4: tap0: <NO-CARRIER,BROADCAST,MULTICAST,UP> mtu 1500 qdisc fq_
codel state DOWN mode DEFAULT group default qlen 1000
    link/ether b2:2e:f2:67:48:ff brd ff:ff:ff:ff:ff:ff
```

TAP interfaces are preferable to use compared to TUN interfaces, as they work at layer 2 and receive and send packets like real Ethernet interfaces.

Now, let's see how we can use veth interfaces.

Using veth interfaces

The network in a Linux container is isolated and has a namespace number associated with it. To connect to them, you will need to use veth interfaces. veth interfaces can be associated with the namespace and can be created on their own or with a peer in a point-to-point configuration. When creating a veth with a peer, you will need to associate the two namespaces, one for each side of the veth peering. Once the peer has been set, any information that is written to one side of the veth peer will be sent to the other side, which is a fast and easy way to interconnect emulated routers when using Linux containers. We are going to use them a lot in our examples. Here is an example of how to create a veth peer interface:

```
claus@dev:~$ sudo ip link add A type veth peer name B
claus@dev:~$ sudo ip link set A netns 41784
claus@dev:~$ sudo ip link set B netns 41634

claus@dev:~$ sudo nsenter -t 41784 -n ip link show A
11: A@if10: <BROADCAST,MULTICAST> mtu 1500 qdisc noop state
DOWN mode DEFAULT group default qlen 1000
    link/ether 9a:fa:1e:7f:0c:34 brd ff:ff:ff:ff:ff:ff link-
netnsid 1
claus@dev:~$ sudo nsenter -t 41634 -n ip link show B
10: B@if11: <BROADCAST,MULTICAST> mtu 1500 qdisc noop state
DOWN mode DEFAULT group default qlen 1000
    link/ether d6:de:78:9c:e9:73 brd ff:ff:ff:ff:ff:ff link-
netnsid 1
```

In this example, two containers are being used, which are identified by the 41784 and 41634 network namespaces. A peer is created with interface names A and B, but the communication between the containers will only be possible after associating the interface name with the network namespace using ip link set <ifname> netns <namespace>, as in this example. The interface names can be the same but need to be renamed only after being associated with the namespace. This is because, before the association, the veth interfaces are in the host, and therefore in the same namespace, which won't allow multiple interfaces with the same name to be created.

Now, let's learn how we can use software bridges.

Using software bridges

Software bridges are used to interconnect software and hardware network ports, which can be added and removed like a real network switch. The Linux kernel has a native software bridge that can be used by using the bridge command or by adding the bridge-utils package and using the brctl command. When a software bridge is created, it requires a name that is also attributed to a network interface that will or will not have an IP address. The following is an example of creating a bridge and associating three interfaces with it:

```
claus@dev:~$ sudo brctl addbr Mybridge
claus@dev:~$ sudo brctl addif Mybridge tap0
claus@dev:~$ sudo brctl addif Mybridge tap1
claus@dev:~$ sudo brctl addif Mybridge enp0s3

claus@dev-sdwan:~$ brctl show Mybridge
bridge name     bridge id      STP    enabled     interfaces
Mybridge        8000.f65..     no                 enp0s3
                                                   tap0
                                                   tap1
```

As explained previously, the `Mybridge` bridge is also associated with a network interface on the Linux host. This can be seen by running the `ip link` command, as shown here:

```
claus@dev-sdwan:~$ ip link show Mybridge
12: Mybridge: <NO-CARRIER,BROADCAST,MULTICAST,UP> mtu 1500
qdisc noqueue state DOWN mode DEFAULT group default qlen 1000
    link/ether f6:78:c6:1a:1c:65 brd ff:ff:ff:ff:ff:ff
```

Linux native bridges are fast and simple, but there are advanced configurations that cannot be performed by them. To be able to use more advanced commands, the recommendation is to use **OpenvSwitch**, also known as **OvS** (there are more details at `https://www.openvswitch.org/`).

Using VXLAN

Linux bridges, TAP, and veth interfaces are used locally inside the Linux host to create a connection between emulated routers, but it won't work to interconnect emulated routers that are running in a different host. There are a few techniques that could be used to connect emulated routers across hosts, such as pseudowires, L2TP, and layer 2 VPNs, among others, but the best would be VXLAN.

VXLAN works as a layer 2 tunnel that extends the local bridge to another remote device, which can be another Linux host, a network switch, or a router. With VXLAN, it is also possible to connect emulated routers to real routers as if they are connected by a wire on a back-to-back connection. As we are going to see later in this section, VXLAN is going to be used for hybrid labs where a connection between real routers and emulated routers is used.

VLAN is a well-known protocol and was explained in *Chapter 2*, Original VLANs have 12 bits of identification, which allows for up to 4,096 VLAN IDs. But VLAN tagging (IEEE 802.1Q) adds an extra 12 bits, so that's up to 24 bits for identifying a VLAN on a normal Ethernet layer 2 frame using VLAN tagging.

VXLAN is independent of VLAN or VLAN tagging, using a header identifier with 24 bits and UDP as transport with port number 4789. An original Ethernet frame that uses a VXLAN tunnel will need an extra 54 bytes of overhead for the tunnel. So, if your network MTU is 1,500 bytes, the maximum MTU of payload that can be carried inside the tunnel will be reduced by 54 bytes. It is recommended to increase MTU when using VXLAN. The following diagram shows an example of protocol encapsulation for VXLAN:

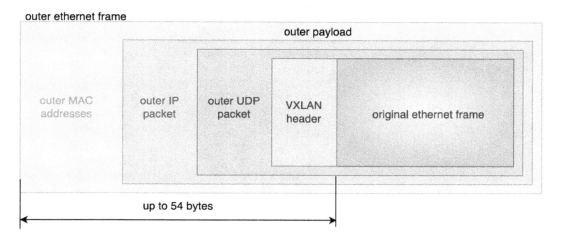

Figure 9.5 – VXLAN encapsulation

A VXLAN connects endpoints, known as **Virtual Tunnel Endpoints** (**VTEPs**). A device that works with VXLAN when receiving an Ethernet frame on a VTEP will then add the VXLAN header, UDP, and IP and will send it toward the other VTEP destination.

Now, let's see how we would set up a hybrid lab using software connections and physical wires.

Building a hybrid lab

A hybrid lab is necessary whenever you want to combine emulated routers with physical routers. The main reason you will need such a configuration is when testing the performance of a real router and adding complexity such as an extra 500 routers in the OSPF backbone. It is also useful to test connections with different complex topologies, such as an external BGP with flapping routes. All extra anomalies can be automated and added using the emulated environment, helping the test gain agility and accuracy.

With a hybrid lab, you could have a few real routers connected to an unlimited number of emulated routers, perhaps building an entire network emulation that can be connected to real routers for closer production environment testing. Again, anomalies can be easily introduced automatically in the emulation, with precision, including packet loss, latency, and jitter. Therefore, your network automation skills are going to be the key to the success of a hybrid lab.

The following diagram shows an example of a hybrid lab connecting four emulated routers to two real routers:

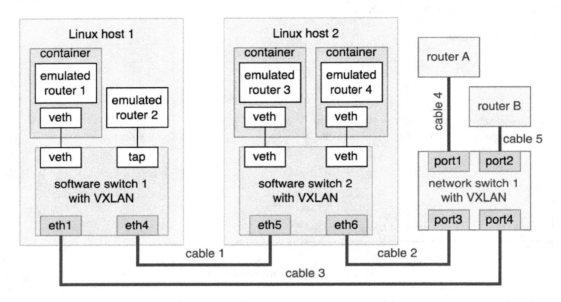

Figure 9.6 – Connectivity example for a hybrid lab

Note that in the preceding figure, the purple lines represent the physical cables that connect Linux host 1, Linux host 2, router A, and router B to network switch 1. VXLAN is used between these devices to allow the setup of any connection between these devices. Linux host 2 uses only container-emulated routers, and therefore veth interfaces. Linux host 1 uses a tap interface to connect emulated router 2, which could be, for instance, Dynamips with a Cisco-emulated router.

The following diagram shows a more complex hybrid setup:

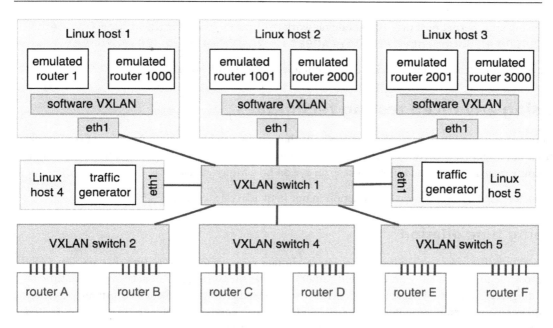

Figure 9.7 – A more complex hybrid lab setup

As you can see in the preceding figure, there are 3,000 emulated routers and 6 physical routers connected via VXLAN switches and software VXLAN bridges. The setup looks clean and straightforward, but it can create very complex connections and topologies. The VXLAN switches work as patch panels that can be configured by software. It is also necessary to have a real router with all interfaces connected to a VXLAN switch so that it can work as a patch panel for that router, such as router A connected to VXLAN switch 2.

Now, let's discuss how we can add an OOB network.

Adding an OOB network to your lab

One important problem we need to pay attention to is how we access devices without needing to have any network connection. And the answer to this problem is to use an OOB network, or out-of-band management network, as we discussed in *Chapter 1*.

Adding some sort of access to devices that do not require any network connection helps when performing catastrophic tests, such as when routers must be removed or turned off. There are several ways to access an emulated router, which is done by accessing the host where the emulation is running. For real routers, the way to access them is via a console or auxiliary ports, which is normally performed using serial communication. Therefore, to allow automation for all devices, you will require a serial port server device that will allow remote access via IP and SSH. Once connected to the serial port server via SSH, you will be able to access the router via a serial port from the port server. One example of this port server is the product Avocent ACS8000, which has 32 serial ports and can be accessed via

IP using an Ethernet port or a 4G mobile network (`https://www.amazon.com/Avocent-ACS8000-Management-Cellular-ACS8032-NA-DAC-400/dp/B099XFB39R`).

Now, let's use some advanced techniques to enhance our network code testing.

Using advanced testing techniques

I created this section to explore some methods that can be used for testing that are not commonly used but might be somehow useful. These techniques are perhaps not used much today but might become mainstream in the future, so keep an eye on how things evolve.

First, let's see how we can use time dilation in our network code testing.

Using time dilation

When building your test environment, you might face requirements for testing that are physically impossible to do in a lab with emulated routers, such as measuring protocol convergence time or sending large amounts of data between devices. The reason these high-performance tests are physically impossible using emulation is that the CPU and I/O on an emulated router are smaller and limited compared to real routers. One way to overcome this limitation is to use **time dilation**.

Time dilation is a technique that changes the CPU clock of the emulated environment in such a way that the emulated router will run slower compared to an emulation without time dilation. From the host's perspective, the emulated router with time dilation will use fewer resources, as it is not running as fast as the emulated router without time dilation. But from the emulated router's perspective, which uses time dilation, everything seems to run at normal speed, but in reality, is much slower.

Imagine you want to test the time it takes for an application to copy files between two ends using network emulation. In a real network, these devices will have 10 GE interfaces that can have up to 10 Gbps. But in an emulation environment, they might have only 100 Mbps available or even less. To overcome these limitations, one technique is to put all emulations, including the application, in a time dilation environment that has a **time dilation factor** (or **TDF**) of 1,000 or more. With a TDF of 1,000, the CPU and I/O, including network interfaces, will be capable of performing more work from the emulated network and application perspective.

For network testing, normally, TDF is greater than 1, but it is also possible to use one smaller than 1, meaning that the emulation will run faster than the host. The applications that use lower than 1 TDF are normally used when tests need to go faster and CPU constraints do not exist. For example, a TDF of 0.1 will run 10 times faster, but the implementation is normally not trivial and sometimes not possible because it relies on shortening the waiting time.

The University of San Diego has done some work on time dilation based on a paper called *To Infinity and Beyond: Time-Warped Network Emulation*, which does have some implementation code for Xen and Linux kernel. Details can be found at `https://www.sysnet.ucsd.edu/projects/time-dilation`. Other implementations for time dilation focus on virtual machines and use Qemu to manipulate time; one implementation is from the University of North Carolina, where they created a project called *Adaptive Time Dilation*. Details can be found at `https://research.ece.ncsu.edu/wireless/MadeInWALAN/AdaptiveTimeDilation`.

One advantage of using time dilation is that TDF can be adjusted every time you perform a test in your environment. A low TDF will impose more CPU and I/O limitations and might be useful to test how the application and network would perform in low-performance conditions, giving a lower-bound test result. A high TDF would give enough CPU and I/O resources to test the application and network in an ideal world without resource limitations, giving an upper-bound test result. Adjusting TDF to a certain mid value would give you the resources compatible with a real network with a real application, giving a test result that's closer to reality.

Now, let's see how we can use monkey testing on network code testing.

Using monkey testing

In software development, monkey testing is used when an application requires testing and there is no formal definition of what to test. Monkey testing generates random inputs and evaluates the outputs.

The idea of monkey testing is to find problems with the code that were not expected by the software developer. In our case, we can use monkey testing to evaluate the network automation tools and the network itself. In some sense, penetration testing would be considered part of monkey testing, but more related to finding security holes, not failures.

One great example of monkey testing is the Netflix implementation called **Chaos Monkey** (there are more details at `https://github.com/Netflix/chaosmonkey`). Chaos Monkey is written in Go and used for randomly terminating containers and virtual machines. With a few modifications, it can be used to shut down interfaces and network links. It is based on chaos engineering (there are more details at `http://principlesofchaos.org/`). Chaos Monkey is a good example of monkey testing that can be added to our network automation testing infrastructure.

Now, let's explore how we could use **artificial intelligence** (**AI**) for network code testing.

Using artificial intelligence

Products that use AI for software development are in their early stages, but they will become reality soon. Today, it is possible to generate code via AI platforms such as **ChatGPT** (there's more at `https://chat.openai.com/auth/login`). Normally, the code generated by the AI is simple and not necessarily optimized, but it is an important milestone toward having to AI write code. The following is an example of the output of using ChatGPT to write unit tests:

Question: Write a unit test for the following code:

```
from paramiko import SSHClient

client = SSHClient()
client.connect('10.1.1.1', username='user', password='pw')
```

AI: Here is the answer:

```python
import unittest
import mock
from paramiko import SSHClient

class TestSSHClient(unittest.TestCase):
    @mock.patch('paramiko.SSHClient.connect')
    def test_connect(self, mock_connect):
        client = SSHClient()
        client.connect('10.1.1.1', username='user',
password='pw')
        mock_connect.assert_called_with('10.1.1.1',
username='user', password='pw')
        client.close()

if __name__ == '__main__':
    unittest.main()
```

Note that the question was very simple and did not specify the language or packages, such as Python or to use the `unittest` and `mock` packages. The AI figured out the language and the packages to import by itself. However, it introduced a small outdated import. Since Python 3.3, `mock` has to be imported as `from unittest import mock`, instead of `import mock`.

Although the unit test written previously was not complete and contained a small error, it is a huge step toward having computers automate tests during development. It can be used today to help developers improve their code.

Some commercial products are already offering AI solutions for software development. For instance, GitHub has a product called **Copilot** that offers pair programming with AI (there are more details at `github.com/features/copilot`). Copilot claims that it can write unit tests for you, which is an amazing achievement.

More and more companies will start offering solutions for code development, and for sure writing unit tests will be the first milestone to be accomplished by AI platforms. Unit tests consume a lot of the developer's time, in most cases taking even more time to write than the code itself. Keep an eye on the market for network automation test tools that use AI – it will make development more robust and faster.

Now, let's see how we can add network simulation to enhance network code testing.

Using network simulation

Network simulation, in contrast to network emulation, uses software to simulate part of the behavior of the network. Most network simulators are used either to simulate network protocol behaviors or to predict and calculate traffic demands and network paths. It can also be used to calculate resources, such as memory and network capacity on devices, but not much more beyond that.

One of the very popular Python packages that is used for network simulation is **NetworkX** (there are more details at `https://networkx.org/`), which is a graph manipulation library. With NetworkX, it is possible to create a large network with thousands of nodes and millions of links using much fewer resources than when using network emulation. Simulating a large network using NetworkX is possible if you wish to run several tests that will be performed much faster than using emulation. However, the tests will evaluate the behavior of the network due to link and node failures, not the control plane (routing protocols) or the operating system of the routers.

Another useful application of network simulation is to test a network access list table path for a particular IP prefix. Once the network simulation has been built, it is possible to determine where a certain IP packet would flow in terms of normal and failure conditions. The simulation must be built using the network router configuration tables, and perhaps can be updated regularly to match production. Note that this kind of test will demand the creation of a network graph for each IP prefix to be tested, and the access list for each interface will dictate the inclusion (or not) of the link in the IP prefix graph.

The following figure shows the topology that we are going to build using NetworkX as an example:

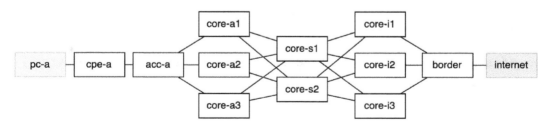

Figure 9.8 – Topology to be used with NetworkX

This topology is also described in the `Chapter09/NetworkX/topology.yaml` file, which is included in the GitHub repo of this book. The following code reads this file and creates a NetworkX graph with this topology:

```python
import networkx as nx
import yaml


G = nx.Graph()
devices = {}


with open("topology.yaml", "r") as file:
    yfile = yaml.safe_load(file)


for i, x in enumerate(yfile["devices"]):
    devices[x] = i
    G.add_node(i, name=x)


for link in yfile["links"]:
    G.add_edge(devices[link[0]], devices[link[1]])
```

After loading the topology, a series of tests can be done to evaluate the behavior of the network. For instance, we could remove the link between `cpe-a` and `acc-a` and see whether there is connectivity between `pc-a` and `internet`. As the process of adding and removing edges is more interactive, the best platform to use NetworkX for testing would be a Jupyter notebook (as described in *Chapter 6*). The following screenshot shows the output of the Jupyter notebook, showing the test of removing a link and testing connectivity between `pc-a` and `internet`:

```
In [48]: nx.node_connectivity(G, devices["pc-a"], devices["internet"])
Out[48]: 1

In [49]: G.remove_edge(1,2)

In [50]: nx.node_connectivity(G, devices["pc-a"], devices["internet"])
Out[50]: 0
```

Figure 9.9 – Jupyter notebook output example

As you can see, if you remove the link between `cpe-a` and `acc-a` (edge `1,2`), `pc-a` loses connectivity to `internet`. The `node_connectivity()` method returns an `integer` that, if greater than zero, indicates there is connectivity between nodes (more details on this method and other connectivity algorithms can be found at `https://networkx.org/documentation/stable/reference/algorithms/connectivity.html`). A series of additional tests can be found in the `Chapter09/NetworkX/example.ipynb` file.

A combination of network simulation and network emulation can be used to increase the capacity and speed of your code testing. A mechanism must be included to build the emulation and create the simulation using the same configuration. In addition, some tests can be performed first on simulation and, if required, repeated in the emulation to be validated.

Using traffic control

Using traffic shaping (or traffic control), it is possible to add complexity to our emulation by adding some physical characteristics that are present in real links and multi-to-multi-point networks. With traffic shaping, we can add latency to specific connections, introduce packet loss, add random limitations, add network congestion, add jitter, and much more. On Linux, it can be easily obtained by using the built-in **Linux Traffic Control** (**TC**), which can be implemented by using the Linux `tc` command.

TC is implemented in Linux by using schedulers or **queuing disciplines** (**qdiscs**). Some qdiscs are included in the Linux kernel, whereas others have to be added as modules. There are classful qdiscs and classless qdiscs, and the difference between them is that one is hierarchical and uses classes and the other does not have any classes (more details on `tc` and qdiscs can be found at `https://tldp.org/HOWTO/Traffic-Control-HOWTO`).

The following is an example of using a qdisc called `netem` to add 10 ms of latency to the loopback interface:

```
$ sudo fping -q -c 100 -p 100 localhost
localhost : xmt/rcv/%loss = 100/100/0%, min/avg/max =
0.030/0.044/0.069

$ sudo tc qdisc add dev lo root netem delay 10ms

$ sudo fping -q -c 100 -p 100 localhost
localhost : xmt/rcv/%loss = 100/100/0%, min/avg/max =
20.4/21.4/24.8
```

This example adds 10 ms to reach the `lo` interface in each way, so the round-trip time is double, which appears in the results as a 21.4 ms average.

Here is another example, showing how to add 5% packet loss using `netem`:

```
$ sudo fping -q -c 100 -p 100 localhost
localhost : xmt/rcv/%loss = 100/100/0%, min/avg/max =
0.031/0.044/0.101


$ sudo tc qdisc add dev lo root netem loss 5%


$ sudo fping -q -c 100 -p 100 localhost
localhost : xmt/rcv/%loss = 100/96/4%, min/avg/max =
0.032/0.056/0.197
```

In this test example, the result was 4% of packet loss instead of 5% as configured. This is because `netem` uses random selection to obtain packet loss, and it will require a larger testing sample to get closer to 5% – for example, 1,000 packets instead of 100, which was used in the preceding test.

Other more complex network behaviors can be added to the interface using `netem`, such as burst control, maximum capacity, network congestion, and random latency variance, among others. More details on it can be found at `https://wiki.linuxfoundation.org/networking/netem`.

There are lots of other schedulers besides `netem`, such as `choke`, `codel`, `hhf`, and ATM. A list of all classless and classful qdiscs available can be obtained on the `tc` man page, which can be visualized by just typing `man tc` (the HTML version can be found at `https://manpages.debian.org/buster/iproute2/tc.8.en.html`).

Hopefully, you have got the most out of this section and have started wondering whether some of these advanced techniques could be added to your project. Adding one of these techniques will likely make your project more reliable and closer to a real production environment.

Summary

The goal of this chapter was to introduce you to how to build and use a proper infrastructure to test your automation code using software. You learned how software can be used to effectively test your automation code, how to use simulation and emulation, how to connect real and emulated devices, and finally, how advanced techniques can be incorporated.

Adding some of the techniques described in this chapter will add superpowers to your network automation code project. From now on, it will be unbeatable.

In the next chapter, we are going to get hands-on in a network lab and wrap up this book with a few additional remarks.

10

Hands-On and Going Forward

Congratulations, you have reached the last chapter of this book, and nothing is better than having some real examples using network automation to help consolidate all of the knowledge learned. We probably won't be able to write examples on all the subjects covered in this book, but the idea is to have at least a foundation for further experimentation and learning.

In this chapter, we are going to build a network from scratch using our network automation skills and emulated routers. The finished network emulated will have enough components for us to experiment with several techniques described in this book. You will be able to use it for your own experimentation whenever you need it.

We are also going to add a few remarks and some guidance for future studies and work, which should be good enough to wrap up the book.

At the end of this chapter, you are going to be able to build your own network emulation and experiment with your own network automation projects within your own computer, which will give you a great environment foundation for future experimentation and learning.

We are going to cover the following topics in this chapter:

- Using a network lab
- Building our network lab
- Connecting the devices
- Adding automation
- Going forward and further studies

Technical requirements

The source code described in this chapter is stored in the GitHub repository at `https://github.com/PacktPublishing/Network-Programming-and-Automation-Essentials/tree/main/Chapter10`.

Using a network lab

It's time for us to do some real code automation tests in a virtual environment. Certain open source and commercial products can be used in your environment for testing network automation. The differences between the commercial and open source solutions rely on the number of different types of devices that are supported and how to scale. Using an open source solution, you might be able to scale up thousands of devices, but it would be limited in terms of the types of devices that can be emulated. A combination of a commercial and open source network might be more useful.

Cisco has a program that allows the public to access emulated routers in a virtual lab; they are called sandboxes. Cisco offers free 24/7 remote access to its sandboxes, but the number of devices is limited. More on Cisco sandboxes can be found at `https://developer.cisco.com/site/sandbox/`.

As an example, the `scrapligo` project, described in *Chapter 6*, uses Cisco sandboxes; check details on the usage at `https://github.com/scrapli/scrapligo/blob/main/examples/generic_driver/interactive_prompts/main.go#L14`. There are also other commercial products, such as Cisco Packet Tracer, which is part of the Cisco Network Academy (`https://www.netacad.com/courses/packet-tracer`), and EVE-NG (`https://www.eve-ng.net/`).

In terms of open source, the most popular ones are GNS3 (`https://gns3.com/`) and Mininet (`http://mininet.org/`). Mininet uses Linux containers to scale up the network, while GNS3 is more focused on virtual machines such as Dynamips. Therefore, GNS3 can run with several different router types but is limited in scale. On the other hand, Mininet can scale to thousands, but with only one router type, which is more appropriate for testing network concepts and network topologies instead of functionalities.

For us, it will be more interesting to start from the ground and build our own network lab, which will give us more insights on how to use it and how we can use network automation tools, which, in fact, will be helpful later for real networks. The basis of our network lab will be Linux containers. Consequently, we need to use routers that can be easily containerized, and because of license issues, we should stick with open source solutions such as FRRouting, Quagga, OpenWRT, or DD-WRT, as described in *Chapter 9*.

Our network lab will use FRRouting as the basis of our routers, which has an interface configuration close to Cisco routers and can be accessed via the `vtysh` command. More details on FRRouting setup and configuration can be found at `https://docs.frrouting.org/en/latest/basic.html`.

Let's now build our own network lab.

Building our network lab

In our network lab, we are going to use Linux containers for all our devices. There will basically be two types of devices, one running a router and one running Linux. The Linux containers that are not working as routers are going to be used to generate traffic or to receive traffic; they are going to mimic a user's PC and a server on the internet.

The intended topology is described in the following diagram:

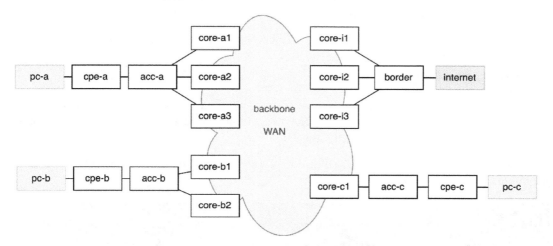

Figure 10.1 – Network lab topology

The containers that are going to work as routers are the white rectangles in *Figure 10.1*, the yellow rectangles are going to work as user PCs, and the green rectangle is going to emulate the servers on the internet.

In total, the network lab will have 16 routers, 3 PCs, and 1 server. The container images to be used in the network lab were created using Docker and stored in Docker Hub (https://hub.docker.com/), which are publicly available and can be used for image download. The routers were created based on the FRRouting docker image (https://hub.docker.com/r/frrouting/frr) and the PCs and server were created based on the Alpine Linux Docker image (https://hub.docker.com/_/alpine).

The original images were slightly modified with a few more tools and configuration changes to create three new images. The image for the routers is hub.docker.com/r/brnuts/routerlab, the image for the PCs is hub.docker.com/r/brnuts/pclab, and the image for the internet is hub.docker.com/r/brnuts/internetlab.

Let's see now how we can launch our lab host.

Launching the lab host

The Linux containers need a host to run from. Therefore, you will first need to launch the Linux host where the routers will be running. I prepared two pre-built images to help, one for VirtualBox and another for Qemu. You can get instructions on how to download them and launch them on GitHub: Chapter10/NetworkLab/README.md. These virtual machine images use Debian Linux.

However, if you don't want to use the pre-built virtual machine, I have also included instructions on how to build your own host, which is basically any Linux distribution with additional packages and some configuration changes. If you build your own image, you will need to start all containers by yourself. I have added a script in Shell that should be able to do that, called `start-containers.sh`.

Once you have launched the host, let's see how we can check whether it was launched properly.

Checking the lab host

After uncompressing and launching the pre-built image, you should be able to see all devices running once the host has finished the boot sequence. The reason is that I have updated the containers so that they restart automatically unless explicitly stopped.

To verify whether the containers representing the devices are running, you just need to use the `docker ps` command, as in the following screenshot:

```
netlab@netlab:~$ docker ps
CONTAINER ID   IMAGE                  COMMAND                CREATED        STATUS        PORTS     NAMES
889847322f65   brnuts/internetlab     "sh -c /etc/docker/d…" 45 hours ago   Up 45 hours             internet
f0c6f2cbc055   brnuts/pclab           "sh -c /etc/docker/d…" 45 hours ago   Up 45 hours             pc-c
13ea431b4870   brnuts/pclab           "sh -c /etc/docker/d…" 45 hours ago   Up 45 hours             pc-b
acfd3fdd7143   brnuts/pclab           "sh -c /etc/docker/d…" 45 hours ago   Up 45 hours             pc-a
eddc0ba95444   brnuts/routerlab       "/sbin/tini -- sh -c…" 45 hours ago   Up 45 hours             cpe-c
8c600215bf9c   brnuts/routerlab       "/sbin/tini -- sh -c…" 45 hours ago   Up 45 hours             cpe-b
cad8f691e3f3   brnuts/routerlab       "/sbin/tini -- sh -c…" 45 hours ago   Up 45 hours             cpe-a
b11f90be9b13   brnuts/routerlab       "/sbin/tini -- sh -c…" 45 hours ago   Up 45 hours             border
b9b9e121b305   brnuts/routerlab       "/sbin/tini -- sh -c…" 45 hours ago   Up 45 hours             acc-c
66acfa649c01   brnuts/routerlab       "/sbin/tini -- sh -c…" 45 hours ago   Up 45 hours             acc-b
0b167e9bb5c8   brnuts/routerlab       "/sbin/tini -- sh -c…" 45 hours ago   Up 45 hours             acc-a
9de3014e0bde   brnuts/routerlab       "/sbin/tini -- sh -c…" 45 hours ago   Up 45 hours             core-i3
94ee0abdf921   brnuts/routerlab       "/sbin/tini -- sh -c…" 45 hours ago   Up 45 hours             core-i2
d26f5046a1f5   brnuts/routerlab       "/sbin/tini -- sh -c…" 45 hours ago   Up 45 hours             core-i1
9346d6dc4d49   brnuts/routerlab       "/sbin/tini -- sh -c…" 45 hours ago   Up 45 hours             core-c1
e757a14e01bf   brnuts/routerlab       "/sbin/tini -- sh -c…" 45 hours ago   Up 45 hours             core-b2
29126eafe6e5   brnuts/routerlab       "/sbin/tini -- sh -c…" 45 hours ago   Up 45 hours             core-b1
c44725e6551d   brnuts/routerlab       "/sbin/tini -- sh -c…" 45 hours ago   Up 45 hours             core-a3
f794b83a68e0   brnuts/routerlab       "/sbin/tini -- sh -c…" 45 hours ago   Up 45 hours             core-a2
08b9989f6498   brnuts/routerlab       "/sbin/tini -- sh -c…" 45 hours ago   Up 45 hours             core-a1
netlab@netlab:~$
```

Figure 10.2 – Output showing all devices running on the network lab

The output of the `docker ps` command should show all running devices, which should be, in total, 20 containers, 16 representing routers (using the `brnuts/routerlab` image), 3 representing PCs (using the `brnuts/pclab` image), and 1 representing the internet (using the `brnuts/internetlab` image).

I have also added volumes to all containers and attached them as persistent storage, so configuration changes won't be removed even on restarting the container. To see the volumes, you can just type in `docker volume list`.

Now, check whether /etc/hosts was updated with the IPs of the containers. You should be able to see several lines after # BEGIN DOCKER CONTAINERS, as in the example in the following screenshot. This file is updated by the update-hosts.sh script included by systemctl:

```
netlab@netlab:~$ grep BEGIN -A 10 /etc/hosts
# BEGIN DOCKER CONTAINERS
172.17.0.10 internet
172.17.0.8 pc-c
172.17.0.12 pc-b
172.17.0.13 pc-a
172.17.0.3 cpe-c
172.17.0.5 cpe-b
172.17.0.17 cpe-a
172.17.0.11 border
172.17.0.4 acc-c
172.17.0.20 acc-b
netlab@netlab:~$
```

Figure 10.3 – Checking the /etc/hosts file

We are going to explain later why we need LLDP in the next section, but for now, let's just check whether lldpd is running on the host using the systemctl status lldpd.service command:

```
netlab@netlab:~$ systemctl status lldpd.service
● lldpd.service - LLDP daemon
     Loaded: loaded (/lib/systemd/system/lldpd.service; enabled; vendor preset: enabled)
     Active: active (running) since Thu 2023-02-02 13:57:26 EST; 5h 44min ago
       Docs: man:lldpd(8)
   Main PID: 325 (lldpd)
      Tasks: 2 (limit: 2337)
     Memory: 8.0M
        CPU: 1.211s
     CGroup: /system.slice/lldpd.service
             ├─325 lldpd: monitor.
             └─365 lldpd: 20 neighbors.

Warning: some journal files were not opened due to insufficient permissions.
netlab@netlab:~$
```

Figure 10.4 – Checking whether lldpd is running on the host

If the lldpd daemon is running correctly, you should be able to see active (running) in green, as in the preceding screenshot.

Now, we should be ready to start doing some network automation to finish building our lab. Let's now see how we connect the devices.

Connecting the devices

Connecting the devices in our network lab will be done by using `veth` peer interfaces as was explained in *Chapter 9*. If we need to connect two different labs from two different hosts, we can use VXLAN, but for our exercise in this section, we are only making connections on the same host. Therefore, `veth` peer interfaces will do the job.

One protocol that I have included in the pre-built virtual machine image and will be very important to us is **Link Layer Discovery Protocol** (**LLDP**). LLDP is one IETF standard that came after the successful Cisco proprietary protocol called **Cisco Discovery Protocol** (**CDP**). It is used to obtain information about the other side of a layer 2 connection by sending specific Ethernet frames. We are going to use it to validate the connections between devices in our network lab.

Before we proceed with our connections, let's check how Docker created our **out-of-band** (**OOB**) management network.

The OOB management network

Docker, by default, creates a network connecting all containers, which we are going to use as our OOB management network (described in *Chapter 1*). To do that, Docker creates `veth` interface peers between the container and the host. On the container side, Docker attributes `eth0` as the name, and on the other side, uses `veth` followed by some hexadecimal characters to make it unique – for instance, `veth089f94f`.

All `veth` interfaces located on the host are then connected to a software bridge called `docker0`. To use the `brctl` command, you might need to install the `bridge-utils` package by doing `sudo apt install bridge-utils`:

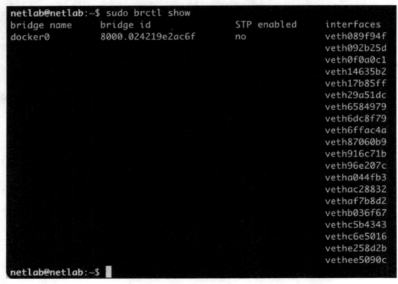

Figure 10.5 – Checking interfaces on the docker0 bridge

To verify which veth interface belongs to which container, you might need to perform two commands, as in the following example:

```
netlab@netlab:~$ docker exec border cat /sys/class/net/eth0/iflink
23
netlab@netlab:~$ grep 23 /sys/class/net/veth*/ifindex
/sys/class/net/veth089f94f/ifindex:23
```

Figure 10.6 – Checking the veth peer name on the host for a container

As you can see in the output of the preceding screenshot, to identify which veth peer interface belongs to the border router, you need to execute a command inside the container to obtain an index for the eth0 interface, which, in this case, was 23. Once you have the index, you can check which veth interface on the host has the index by using a grep command on all files.

We can also use LLDP to find out the veth interface name by doing the command directly on the router:

```
netlab@netlab:~$ docker exec border lldpctl eth0
-------------------------------------------------------------------------
LLDP neighbors:
-------------------------------------------------------------------------
Interface:    eth0, via: LLDP, RID: 1, Time: 0 day, 05:02:59
  Chassis:
    ChassisID:    mac 08:00:27:4d:f8:4f
    SysName:      netlab
    SysDescr:     Debian GNU/Linux 11 (bullseye) Linux 5.10.0-21-amd64 #1 SMP Debian
    MgmtIP:       10.0.2.15
    MgmtIface:    2
    MgmtIP:       fe80::a00:27ff:fe4d:f84f
    MgmtIface:    2
    Capability:   Bridge, on
    Capability:   Router, on
    Capability:   Wlan, off
    Capability:   Station, off
  Port:
    PortID:       mac 2a:8e:f9:b3:1d:bd
    PortDescr:    veth089f94f
    TTL:          120
    PMD autoneg:  supported: no, enabled: no
      MAU oper type: 10GigBaseCX4 - X copper over 8 pair 100-Ohm balanced cable
-------------------------------------------------------------------------
netlab@netlab:~$
```

Figure 10.7 – Showing the LLDP neighbor inside the border router

The preceding screenshot shows a successful output of `lldpctl` showing the `eth0` interface's neighbor, which, in this case, is the host Debian Linux, with a `SysName` of `netlab`. The interface that peers with `eth0` on the border router is described in the `PortDescr` field as `veth089f94f` – the same interface that we discovered using the commands in *Figure 10.6*.

However, why not use the first method described in *Figure 10.6* to find out the connection instead of LLDP? Because in real networks, LLDP is used to identify the connections between devices. Thus, writing an automation code to verify all the network connections in the lab using LLDP can also be used in production. Our lab will be used as the first place to test our automation code – in this case, checking the LLDP topology.

By now, you have probably noticed that we can access the routers just by using the `docker exec <name of the router>` command, so why do we need the OOB management network to access the devices? The answer is like the LLDP case – with OOB network access, the devices can be accessed via SSH, which is what we are going to do in production. Consequently, any code developed for the lab can be used in production.

To test our lab OOB management network, we just need to access the device via the IP using `ping` or `ssh` commands – the `ping cpe-a` command, for example:

```
netlab@netlab:~$ ping -c 5 cpe-a
PING cpe-a (172.17.0.17) 56(84) bytes of data.
64 bytes from cpe-a (172.17.0.17): icmp_seq=1 ttl=64 time=0.083 ms
64 bytes from cpe-a (172.17.0.17): icmp_seq=2 ttl=64 time=0.068 ms
64 bytes from cpe-a (172.17.0.17): icmp_seq=3 ttl=64 time=0.058 ms
64 bytes from cpe-a (172.17.0.17): icmp_seq=4 ttl=64 time=0.067 ms
64 bytes from cpe-a (172.17.0.17): icmp_seq=5 ttl=64 time=0.067 ms

--- cpe-a ping statistics ---
5 packets transmitted, 5 received, 0% packet loss, time 4076ms
rtt min/avg/max/mdev = 0.058/0.068/0.083/0.008 ms
netlab@netlab:~$
```

Figure 10.8 – Testing connection to a router from the host using OOB

You should also be able to SSH to any container, using `netlab` for the username and password:

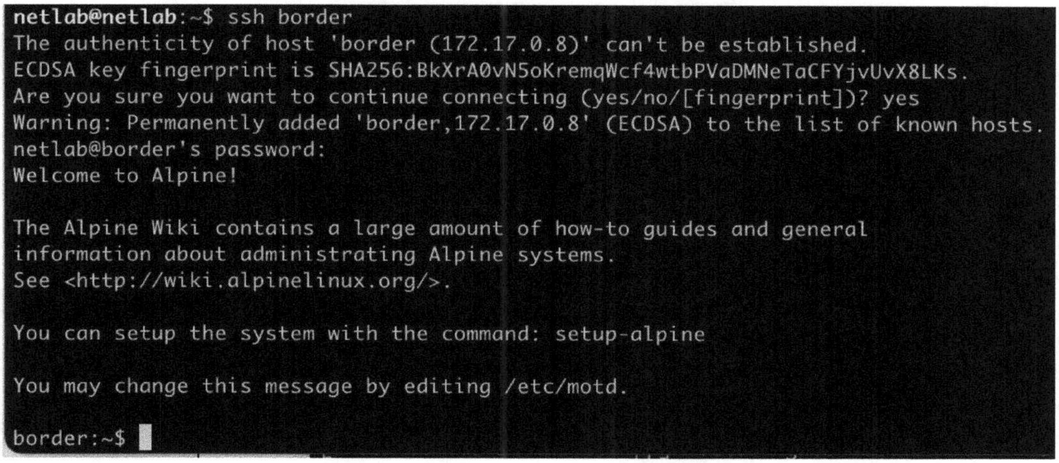

```
netlab@netlab:~$ ssh border
The authenticity of host 'border (172.17.0.8)' can't be established.
ECDSA key fingerprint is SHA256:BkXrA0vN5oKremqWcf4wtbPVaDMNeTaCFYjvUvX8LKs.
Are you sure you want to continue connecting (yes/no/[fingerprint])? yes
Warning: Permanently added 'border,172.17.0.8' (ECDSA) to the list of known hosts.
netlab@border's password:
Welcome to Alpine!

The Alpine Wiki contains a large amount of how-to guides and general
information about administrating Alpine systems.
See <http://wiki.alpinelinux.org/>.

You can setup the system with the command: setup-alpine

You may change this message by editing /etc/motd.

border:~$
```

Figure 10.9 – Testing whether you can access a device via SSH using OOB

Now that we know how our OOB management network works in our lab, let's connect the devices using the OOB network.

Looking at the topology

Figure 10.1 shows the topology that we are aiming to create. The devices in our lab are running and connected to an OOB network, but they do not have connections like the topology described in *Figure 10.1*.

In addition to the diagram, there is also a formal topology description in a file on GitHub, which can be accessed at `Chapter10/NetworkLab/topology.yaml`. The file describes the routers in the topology and their connections. It is a simple version of network definition in YAML format as we discussed in *Chapter 4*.

The topology file has basically two main keys, `devices` and `links`. These keys should describe the same connections shown in *Figure 10.1*. The following is a sample of the file for the `devices` and `links` keys:

```
devices:
  - name: acc-a
    type: router_acc
    image: brnuts/routerlab
  - name: acc-b
    type: router_acc
    image: brnuts/routerlab
```

```
links:
  - name: [pc, cpe]
    connection: [pc-a, cpe-a]
  - name: [cpe, acc]
    connection: [cpe-a, acc-a]
```

The file should contain all the links that are depicted in *Figure 10.1*. Ideally, the diagram in *Figure 10.1* should be created automatically by a tool reading from `topology.yaml`. In our example, the diagram and the YAML file are the same, but I have built the diagram manually myself, and for any topology change, I need to update the `topology.yaml` file and the diagram. This problem was discussed as well in *Chapter 4*, and the update synchronization between a file and a diagram tends to break as the topology gets more complex. But, for our examples using this small topology, an automated diagram builder is not necessary.

Creating the connections between devices

To connect the devices, like in the topology, we have to use `veth` peer interfaces, and as we discussed in *Chapter 9*, we need the namespace numbers for each side of the peer and the interface names we going to use. Most of the connections in *Figure 10.1* are point-to-point between devices, except for the connections to the **backbone** or **WAN**.

The following diagram shows all the `veth` peers that we must configure; the majority are connected between two containers in a point-to-point configuration. However, the core routers will use, let's say, backbone veth or WAN veth, because they are connected in a multi-to-multi-point environment, similar to a WAN. For that, we are going to use a software bridge in the host to provide connectivity between the backbone veth – latency and packet loss can be added to the bridge interfaces if tests for network degradation are required:

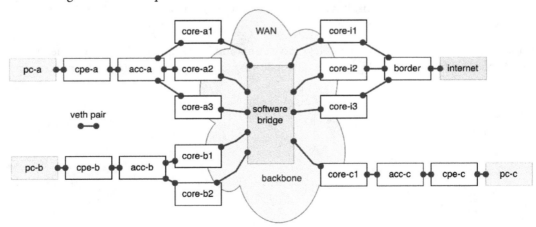

Figure 10.10 – Lab topology showing all veth peer interfaces

When we start creating the veth interfaces for the backbone to connect all core routers, we are going to use one namespace on the host and the other in the core router. This is different from all other veth, which will have two namespaces. The following is one example of how to create a connection manually between pc-a and cpe-a:

```
netlab@netlab:~$ docker inspect -f '{{.State.Pid}}' pc-a
1069
netlab@netlab:~$ docker inspect -f '{{.State.Pid}}' cpe-a
1063
netlab@netlab:~$ sudo ip link add pc-cpe type veth peer name
cpe-pc
netlab@netlab:~$ sudo ip link set pc-cpe netns 1069
netlab@netlab:~$ sudo ip link set cpe-pc netns 1063
netlab@netlab:~$ docker exec pc-a ip link set pc-cpe up
netlab@netlab:~$ docker exec cpe-a ip link set cpe-pc up
```

As we can see in these commands, first, we need to obtain the network namespace IDs of each router we want to connect, and then we can create the veth peer and attribute each side of the peer to a namespace ID. Finally, we bring the interfaces up on each router. Note that the interface name on pc-a is pc-cpe and on cpe-a is cpe-pc, to help identify the direction in which the interface goes.

To verify whether our connection between the routers was created properly, we can run the following command:

```
netlab@netlab:~$ docker exec pc-a lldpctl pc-cpe
-------------------------------------------------------------------------
LLDP neighbors:
-------------------------------------------------------------------------
Interface:     pc-cpe, via: LLDP, RID: 2, Time: 0 day, 00:04:31
  Chassis:
    ChassisID:    mac 02:42:ac:11:00:02
    SysName:      cpe-a
    SysDescr:     Alpine Linux v3.13 Linux 5.10.0-21-amd64 #1 SMP Debian
    MgmtIP:       172.17.0.2
    MgmtIface:    4
    Capability:   Bridge, off
    Capability:   Router, on
    Capability:   Wlan, off
    Capability:   Station, off
  Port:
    PortID:       mac 1e:a6:82:43:4f:b3
    PortDescr:    cpe-pc
    TTL:          120
    PMD autoneg:  supported: no, enabled: no
      MAU oper type: 10GigBaseCX4 - X copper over 8 pair 100-Ohm balanced
-------------------------------------------------------------------------
netlab@netlab:~$
```

Figure 10.11 – Checking the connection between pc-a and cpe-a

Now, we can confirm that `pc-a` is connected to `cpe-a` by looking into the `lldpctl` command output in *Figure 10.11*. The output shows the name for `SysName`, which is `cpe-a`, confirming the connection. We can also see the interface name on the other side, which is `cpe-pc`.

Let's now see how we automate the device connections.

Automating the connections

Our lab now has all devices running, and we are going to connect all devices using a program that will connect all devices. You can get access to the program on `Chapter10/NetworkLab/AUTOMATION.md`.

To install it on your computer, just need to clone it using the following:

```
claus@dev % git clone https://github.com/brnuts/netlab.git
Cloning into 'netlab'…
remote: Enumerating objects: 103, done.
remote: Counting objects: 100% (103/103), done.
```

```
remote: Compressing objects: 100% (79/79), done.
remote: Total 103, reused 44 (delta 12), pack-reused 0
Receiving objects: 100% (103/103), 58 KiB | .9 MiB/s, done.
Resolving deltas: 100% (36/36), done.
```

Then, you need to build the Go program:

```
claus@dev % go build
claus@dev % ls -lah netlab
-rwxr-xr-x  1 user   staff    5.3M Feb  8 11:54 netlab
```

If you are using the pre-build VirtualBox image, you probably are accessing the network lab via SSH on localhost port 22. Then, you just need to run it like so:

```
claus@dev % ./netlab
```

If you are using QEMU or your own Linux virtual machine with the network lab, you can pass the username, password, and IP of the host as follows:

```
claus@dev % ./netlab -host 10.0.4.1 -user oper -pw secret
```

A small help guide can be accessed by adding -help, like here:

```
claus@dev % ./netlab -help
Usage of ./netlab:
  -host string
      Host IP for netlab (default "localhost")
  -port uint
      SSH port to access Host IP for netlab (default 22)
  -pw string
      Password to access netlab host (default "netlab")
  -topo string
      Topology yaml file (default "topology.yaml")
  -user string
      Username to access netlab host (default "netlab")
```

The program shows some logs on the output, and a successful run should show similar lines to the following:

```
claus@dev % ./netlab
2023/02/08 12:22:49 reading topology.yaml file
2023/02/08 12:22:49 connecting via SSH to netlab@localhost
2023/02/08 12:22:49 loading veth information
2023/02/08 12:22:50 connecting devices with 25 veths
2023/02/08 12:23:00 creating bridge and adding veth backbones
2023/02/08 12:23:01 all done successfully
claus@dev %
```

Figure 10.12 – Running the Go automation program to connect devices

As you can see in the preceding screenshot, the program takes around 12 seconds to run, and it should show all done successfully at the end.

Let's have a look at this program and what it is doing.

Looking into the automation program

In our example, the program was written in Go, and the directory where it is located consists of nine files, of which six are Go source code with .go extensions as shown here:

```
claus@dev % ls -1
go.mod
go.sum
hostconnect.go
netlab.go
readtopology.go
runcommand.go
topology.yaml
types.go
vethcommands.go
```

Let's discuss each file.

go.mod and go.sum

These files are used by the Go builder for package dependency management; they contain all the necessary information to add the third-party library to our program. Every time we import a package, it automatically updates these files. More on these files can be obtained at https://go.dev/ref/mod.

topology.yaml

This contains the description of the topology that is shown in *Figure 10.1.*

types.go

This contains all the data structure definitions used in the program, which include variable types and the YAML topology data structure. Different from Python, in Go, it is better to specify the data structure that you are going to read from a YAML file. In our case, the `TopologyConfType` struct type is used to define the YAML file structure, like so:

```go
type DeviceTopologyType struct {
        Name    string
        Type    string
        Image   string
}


type LinkTopologyType struct {
        Name          [] string
        Connection    [] string
}


type TopologyConfType struct {
        Devices   [] DeviceTopologyType
        Links     [] LinkTopologyType
}
```

readtopology.go

This contains the function that is used to read the `topology.yaml` file. The data structure of this file is defined in the `TopologyConfType` structure type.

runcommand.go

This contains generic functions that wrap the command to run on the host of the lab. If an error occurs by running a command, the output is combined with the error message to be returned in the error message, as in this example:

```go
fmt.Errorf("failed '%s', out: %s ,err: %v", cmd, out, err)
```

The idea to add the output to the error message is because when running remote commands via SSH and shell, the error might not be easy to interpret without the `stdout` messages.

veth.go

This contains the functions that form the command strings that will be used to run on the host to create or manipulate `veth` interfaces. It also contains all functions that are used to populate the `conf.Veths` list, such as `loadVeth()` and `createPeerVeths()`.

hostconnect.go

This file contains the function used to connect to our lab. In our case, we are using a third-party package called `melbahja/goph`, which is an SSH client that allows the execution of a command and immediate output. For faster and better performance, we should use vSSH instead, as explained in *Chapter 6*.

netlab.go

This is the main program file that contains the `main()` and `init()` functions. The library called `flags` is used to pass arguments during the command execution in the shell. By default, they are initiated in the `init()` function, and they use default values if the arguments are not passed.

The `main()` function also describes the flow of the whole process, which consists of five main calls – `readTopologyFile`, `connectToHost`, `loadVeths`, `createVeths`, and `addVethsToBackbone`.

Now that we have all devices connected and we understand how the automation works, let's do some manual checks to verify that the connections have been created properly.

Checking the connections manually

To create automation for checking connections, we have to understand how the process of checking connections works first. Once we know how a manual check works, we can later automate it.

Once the `netlab` program has run without errors, it should have created the connections and the software bridge called `backbone`, and attached the WAN interfaces to it. Let's use the following figure as guidance for our manual verification of the connections:

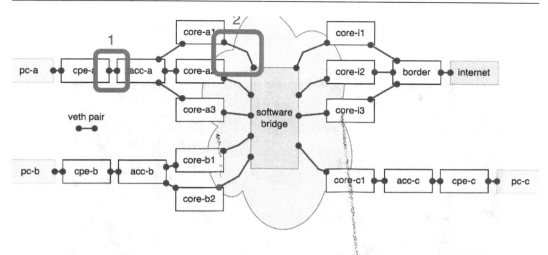

Figure 10.13 – Diagram showing where we manually check the connections

The figure shows numbers to indicate where we are going to do the manual checks. Let's first start checking the connection that is represented by **1**. We are going to use LLDP to validate the connection in all cases.

The following screenshot shows the output of the `sudo lldpctl cpe-acc` command, which runs inside the `cpe-a` router via SSH. Note that in the example, we start from the `netlab` host:

```
netlab:netlab# ssh cpe-a "ash -c 'sudo lldpctl cpe-acc'"
netlab@cpe-a's password:
-------------------------------------------------------------
LLDP neighbors:
-------------------------------------------------------------
Interface:    cpe-acc, via: LLDP, RID: 2, Time: 0 day, 02:25:27
  Chassis:
    ChassisID:    mac 02:42:ac:11:00:0b
    SysName:      acc-a
    SysDescr:     Alpine Linux v3.13 Linux 5.10.0-18-amd64 #1 SMP Debian
    MgmtIP:       172.17.0.11
    MgmtIface:    113
    Capability:   Bridge, off
    Capability:   Router, on
    Capability:   Wlan, off
    Capability:   Station, off
  Port:
    PortID:       mac ce:25:15:ba:4d:90
    PortDescr:    acc-cpe
    TTL:          120
    PMD autoneg:  supported: no, enabled: no
      MAU oper type: 10GigBaseCX4 - X copper over 8 pair 100-Ohm balanced
-------------------------------------------------------------
```

Figure 10.14 – Output of the lldpctl command to verify the cpe-a connection to acc-a

As you can see, the `cpe-acc` interface in the `cpe-a` router is connected to the `acc-cpe` interface in the `acc-a` router.

To validate the connection for case **2** in *Figure 10.14*, we will run LLDP on `core-a1`:

```
netlab:netlab# ssh core-a1 "ash -c 'sudo lldpctl core-a1-wan'"
netlab@core-a1's password:
-------------------------------------------------------------------------------
LLDP neighbors:
-------------------------------------------------------------------------------
Interface:      core-a1-wan, via: LLDP, RID: 1, Time: 0 day, 02:39:43
  Chassis:
    ChassisID:    mac 08:00:27:65:ae:c0
    SysName:      netlab
    SysDescr:     Debian GNU/Linux 11 (bullseye) Linux 5.10.0-18-amd64 #1 SMP Debian
    MgmtIP:       10.0.2.15
    MgmtIface:    2
    MgmtIP:       fe80::a00:27ff:fe65:aec0
    MgmtIface:    2
    Capability:   Bridge, on
    Capability:   Router, on
    Capability:   Wlan, off
    Capability:   Station, off
  Port:
    PortID:       mac 1a:ce:dc:00:74:4e
    PortDescr:    wan-core-a1
    TTL:          120
    PMD autoneg:  supported: no, enabled: no
      MAU oper type: 10GigBaseCX4 - X copper over 8 pair 100-Ohm balanced cable
-------------------------------------------------------------------------------
```

Figure 10.15 – Output of the lldpctl command to verify the core-a1 connection to the backbone

As you can see, the `core-a1` router's interface, `core-a1-wan`, is connected to the `netlab` host via `wan-core-a1`. To verify whether the `wan-core-a1` interface belongs to the `backbone` bridge, we need to perform one of the extra commands:

```
root@netlab:~# ip link show wan-core-a1
167: wan-core-a1@if168: <BROADCAST,MULTICAST,UP,LOWER_UP> mtu 1500 qdisc noqueue
 master backbone state UP mode DEFAULT group default qlen 1000
    link/ether 1a:ce:dc:00:74:4e brd ff:ff:ff:ff:ff:ff link-netnsid 0
root@netlab:~# ip -details -pretty -json link show wan-core-a1 | grep master
      "master": "backbone",
root@netlab:~# 
```

Figure 10.16 – Command to check that wan-core-a1 belongs to the backbone bridge

Either of the commands shown in *Figure 10.16* confirms that wan-core-a1 belongs to `backbone`. The difference is the second command presents the output in a JSON format, which is easier to parse by software. `lldpctl` also supports JSON output using `lldpctl -f json`.

Now, let's discuss how we can add more automation.

Adding automation

There are infinite possibilities for processes for which you might want to create automation. Most of the operational procedures are repetitive and prone to errors if manually operated. So, we need to automate our network as much as possible.

Let's then describe a few simple forms of automation that can help our network operation.

Link connection check automation

One of the procedures that is very important and requires lots of attention is the build and construction of the physical network, in particular a physical rack and its cables. Its complexity will vary, depending on whether a star topology configuration or a Clos topology configuration is used, which we discussed in *Chapter 1*.

A mixed topology configuration that combines all possible topologies is even more complex, and its complexity will increase the chances of building a network incorrectly. For instance, a Clos network, as shown in *Figure 10.17*, has a total of 32 connections, and imagine the complexity added if three more routers were included.

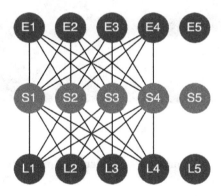

Figure 10.17 – Clos network connections

Having included **E5**, **S5**, and **L5**, the Clos network will have now 50 connections. So, for us, connection check automation is important to avoid operation failures down the network setup.

Most importantly, our network lab can be used for us to test the automation of the link connection check, which can be used later in production.

In a production environment, a bastion host normally needs to be accessed first, before accessing the devices using the OOB network. In our network lab, the bastion is the same as the network lab host. Once logged in to the bastion, the automation code can then access the router via the OOB network, which is the same as in the network lab.

Let's now write some code to automate this process

Link check example code

I have added a Python script that returns a JSON list format with all interfaces and the device connected to each interface for a particular device. The Python code can be accessed at Chapter10/ NetworkLab/AUTOMATION.md.

Let's run a few examples to see how the Python script works; the following screenshot is showing results for the internet device:

```
claus@dev % python3 ./show-connections.py -t internet
[
    {
        "Interface": "eth0",
        "Device": "netlab"
    },
    {
        "Interface": "internet-border",
        "Device": "border"
    }
]
claus@dev %
```

Figure 10.18 – Output for checking connections to the internet device

As you can see, the internet device has only two interfaces, one connected via the OOB network with the eth0 interface to the netlab device, and one interface called internet-border connected to the border device, which confirms the connections in *Figure 10.1*.

Let's now check how the border device is connected.

```
claus@dev % python3 ./show-connections.py -t border
[
    {
        "Interface": "eth0",
        "Device": "netlab"
    },
    {

        "Interface": "border-internet",
        "Device": "internet"
    },
    {

        "Interface": "border-core1",
        "Device": "core-i1"
    },
    {

        "Interface": "border-core2",
        "Device": "core-i2"
    },
    {

        "Interface": "border-core3",
        "Device": "core-i3"
    }
]
claus@dev %
```

Figure 10.19 – Output for checking connections of the border device

As you can see in *Figure 10.19*, the border device is connected to the internet device and three core routers, core-i1, core-i2, and core-i3, as in *Figure 10.1*.

If you run this for all devices, you should confirm all the connections. But can we automate confirming connections for all routers in just one run? Yes, of course, but for that, we will need to read topology. yaml, and then create a loop that will run on each device to confirm the connections. I will leave completing this as an exercise for you.

Let's now explain some parts of the show-connections.py code.

Looking into the code

The show-connections.py Python code uses the paramiko third-party library as the base for the SSH connection, which we discussed in *Chapter 6*, (paramiko can be installed using pip install paramiko).

Paramiko is a lower-level SSH connectivity library that allows us to create an SSH session within an SSH session because we are using a bastion to connect to our devices in the network lab, which is the lab host. The details of this stacked SSH connection are described in the code by the NetLab()

class, which has a method to connect to the bastion called `connectBastion()`, and a method to connect to a device called `connectDevice()`. Note that these methods use a class attribute called `self.transport` to pass the bastion `paramiko` handler to the device channel described in the code and shown here:

```
device_channel = self.transport.open_channel(
    "direct-tcpip", target_socket, source_socket
)
```

There are other ways to use bastion, such as using SSH proxies or SSH agents. However, in our example, I wanted to show how to natively create an SSH stack connection. Because if you do have two bastions before connecting a device, it is also possible to use `paramiko`, but perhaps not that easy using SSH agents and proxies.

In the Python code, we use `argparser` to add arguments to our command line, so you can change the address of the bastion or the username and password. The arguments and default values are located in `parse_arguments()`. A help guide is also automatically produced if you type `--help`.

```
claus@dev % python3 ./show-connections.py --help
usage: show-connections.py [-h] [-b BASTION] [-u B_USER] [-p B_PASSWD]
-t DEVICE [-r R_USER]
                                    [-s R_PASSWD]

options:
  -h, --help            show this help message and exit
  -b BASTION, --bastion BASTION
                        Bastion address to connect
  -u B_USER, --user B_USER
                        Username for the bastion access
  -p B_PASSWD, --pass B_PASSWD
                        Password for the bastion access
  -t DEVICE, --targetdevice DEVICE
                        Target device to find interface neighbours
  -r R_USER, --ruser R_USER
                        Username for the device access
  -s R_PASSWD, --rpass R_PASSWD
                        Password for the device access
claus@dev %
```

Figure 10.20 – help output for the Python code show-connections.py

I will leave you to improve this Python script to read the `topology.yaml` file and then verify all the connections in the network lab depicted in *Figure 10.1*.

Now, let's see how we can automate the IP configuration for the interfaces.

IP configuration automation

Before we can use our network for IP traffic, we need to attribute IPs to our network interfaces, which can be done manually by adding IPs to each interface, or we can create an automated code that distributes the IPs and configure them on the devices in the network lab.

For our network lab, there are basically two types of IP allocation, one that is point-to-point between devices, and one that is multi-point for the WAN on the backbone interfaces. Let's give an example of automation for the WAN interfaces. The Python code called `configure-ip-wan.py` is located at `Chapter10/NetworkLab/AUTOMATION.md`.

The following screenshot shows the output after running the `configure-ip-wan.py` program:

```
claus@dev % python3 configure-ip-wan.py
main(): reading topology file
main(): getting wan interfaces
main(): configuring inteface wan devices with 10.200.200.0/24 subnet
run_device_command(): running cmd 'sudo ip addr add 10.200.200.1/24 dev core-a1-wan' on device 'core-a1'
run_device_command(): running cmd 'sudo ip addr add 10.200.200.2/24 dev core-a2-wan' on device 'core-a2'
run_device_command(): running cmd 'sudo ip addr add 10.200.200.3/24 dev core-a3-wan' on device 'core-a3'
run_device_command(): running cmd 'sudo ip addr add 10.200.200.4/24 dev core-b1-wan' on device 'core-b1'
run_device_command(): running cmd 'sudo ip addr add 10.200.200.5/24 dev core-b2-wan' on device 'core-b2'
run_device_command(): running cmd 'sudo ip addr add 10.200.200.6/24 dev core-c1-wan' on device 'core-c1'
run_device_command(): running cmd 'sudo ip addr add 10.200.200.7/24 dev core-i1-wan' on device 'core-i1'
run_device_command(): running cmd 'sudo ip addr add 10.200.200.8/24 dev core-i2-wan' on device 'core-i2'
run_device_command(): running cmd 'sudo ip addr add 10.200.200.9/24 dev core-i3-wan' on device 'core-i3'
claus@dev %
```

Figure 10.21 – Output of the configure-ip-wan.py Python code

Note that the IPs are configured on the devices using Paramiko, as in the previous example. The code uses the `ipaddress` Python library, which allocates the IPs that will be used in the WAN interfaces by creating a list of IPs using the following commands:

```
network = ipaddress.ip_network(args.subnet)
valid_ips = list(network.hosts())
```

Then, each IP is obtained by using `pop()` in the `valid_ips` list, like in the following loop:

```
prefix = network.prefixlen
for device, interface in wan_interfaces.items():
    ip = valid_ips.pop(0)
    cmd = "sudo ip addr add {}/{} dev {}".format(ip, prefix,
interface)
    run_device_command(args, device, cmd)
```

Now, we can test the IP connectivity between devices in the WAN using the Python script included at `Chapter10/NetworkLab/AUTOMATION.md`:

```
claus@dev % python3 ./run-command.py -t core-a1 -c "ping -c 5 10.200.200.4"
PING 10.200.200.4 (10.200.200.4): 56 data bytes
64 bytes from 10.200.200.4: seq=0 ttl=42 time=0.229 ms
64 bytes from 10.200.200.4: seq=1 ttl=42 time=0.772 ms
64 bytes from 10.200.200.4: seq=2 ttl=42 time=0.175 ms
64 bytes from 10.200.200.4: seq=3 ttl=42 time=0.294 ms
64 bytes from 10.200.200.4: seq=4 ttl=42 time=0.371 ms

--- 10.200.200.4 ping statistics ---
5 packets transmitted, 5 packets received, 0% packet loss
round-trip min/avg/max = 0.175/0.368/0.772 ms
```

Figure 10.22 – Testing the IP connectivity between core-a1 and core-b1 via the WAN

From the output shown in *Figure 10.21*, we can assume the `core-b1` interface IP is `10.200.200.4`. So, the test executed on `core-a1` is testing the IP connectivity between `core-a1` and `core-b1` via the `backbone` bridge.

To finish the IP configuration, you will have to also configure IPs for all other interfaces. I will leave adding the IPs to other network lab interfaces as an exercise, but for now, the example is sufficient to guide you to proceed.

Additional network lab automation

Let's discuss briefly what other possible automation we can add to our lab.

Adding and removing devices

We can add code that can read the `topology.yaml` file and then, based on what is running, determine whether certain modifications are required, such as adding devices or removing devices. I will say that is easier for us to just tear down a network lab and start another one from scratch instead of removing and adding devices because, in our network emulation, the shutdown and startup are quick.

So, adding code to remove and add devices is more of an exercise than a real utility in our network lab.

Using gRPC for automation

We also can do some automation using gRPC, as FRRouting supports this interface. With this, we eliminate the necessity to access the devices via SSH. You can find more on gRPC for FRRouting at `https://docs.frrouting.org/en/latest/grpc.html`.

Using NETCONF for automation

To use NETCONF for automation, you need to have libyang installed in the router image, which, in our network lab, is FRRouting running on Alpine Linux. To add libyang, just type the sudo apk add libyang command on the router device. Using FRRouting and NETCONF together is not a very well-documented option, so good luck doing so.

Adding network degradation

You can add latency, jitter, packet loss, congestion, and other degradations to your network lab, which can be permanent or vary over time. To remove and add these degradations, the best thing to do is to write automated code that can apply the necessary traffic shaping mechanisms and then remove them. We discussed these degradation methods in *Chapter 9*.

As an example, we can add some latency to the backbone interfaces in our network lab by using Linux tc like so:

```
sudo tc qdisc add dev wan-core-a1 root netem delay 100ms
```

This command should run on the lab host, and it will add a 100 ms delay to the wan-core-a1 interface that connects core-a1 to the backbone.

Figure 10.23 show the same test done in *Figure 10.22* but with WAN latency added.

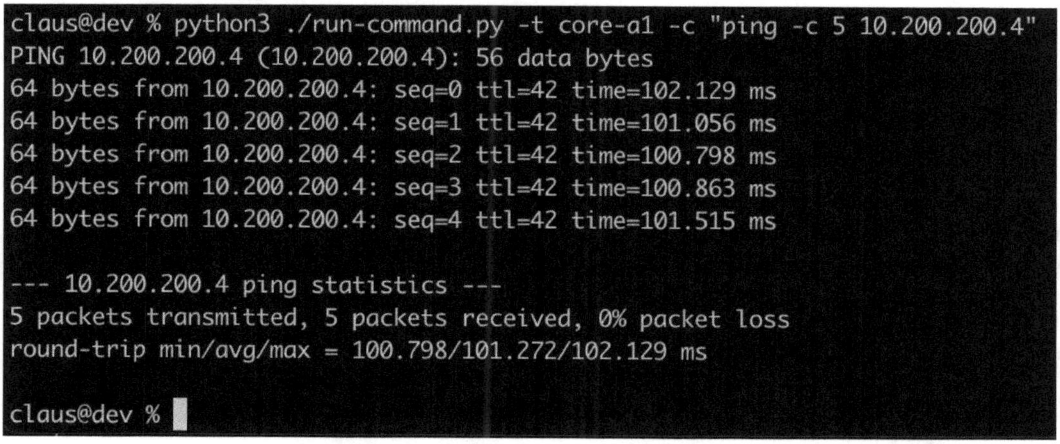

```
claus@dev % python3 ./run-command.py -t core-a1 -c "ping -c 5 10.200.200.4"
PING 10.200.200.4 (10.200.200.4): 56 data bytes
64 bytes from 10.200.200.4: seq=0 ttl=42 time=102.129 ms
64 bytes from 10.200.200.4: seq=1 ttl=42 time=101.056 ms
64 bytes from 10.200.200.4: seq=2 ttl=42 time=100.798 ms
64 bytes from 10.200.200.4: seq=3 ttl=42 time=100.863 ms
64 bytes from 10.200.200.4: seq=4 ttl=42 time=101.515 ms

--- 10.200.200.4 ping statistics ---
5 packets transmitted, 5 packets received, 0% packet loss
round-trip min/avg/max = 100.798/101.272/102.129 ms

claus@dev %
```

Figure 10.23 – Same test done in Figure 10.22 with 100 ms latency added to WAN

Feel free to add other network degradations to your network lab by automating how Linux traffic control can be used in the network lab interfaces.

Configure routing

At this stage, our network lab does not provide IP traffic capabilities because it does not know how to route IP packets. Two kinds of routing can be done using static routes by configuring the interfaces with them or adding dynamic protocol for the devices to talk and exchange routing tables.

As all routers in our network lab are FRRouting; the protocols available that can be used are EIGRP, OSPF, ISIS, RIP, or BGP. A complete list and more details can be found at `https://docs.frrouting.org/`.

Many more kinds of automation are possible. Some will only work for the network lab, but some could be used in a production network. Hopefully, you can use the network lab to improve your network automation code skills, and then gain more confidence on building a solution for a production network.

Let's now discuss what to do next and further study.

Going forward and further study

You are probably now thinking about what to do next and how you can progress to the next topic in network automation. I have put together a few suggestions here that I would recommend following, but keep in mind that there might be many other paths to follow. So, it is just a humble suggestion, and I hope you can enjoy the journey.

Checking popular platforms and tools

There are many automation platforms that you can use and perhaps investigate how they work. This list will be dynamic and may change from one year to another. So, keep in mind how to search for them and how to evaluate them.

I have done superficial research on some of the platforms, but I would recommend going deeper if you want to improve your knowledge and get even sharper on how to automate a network. You may get some ideas, and perhaps improve what you are doing today.

Here is a small list of the most popular automation platforms and tools that you might want to have a look at, not in any particular order:

- The Salt project:
  - Short description: A remote execution manager
  - Source: `https://github.com/saltstack/salt`
  - Contributors: Over 2K
  - Repository creation date: February 2011
  - Top languages: Python 98%

- License: Apache 2.0

- Sponsor: VMware/public

- Popularity: 13K stars, 5.4K forks, 544 watching

• The Ansible project:

  - Short description: A simple automation system for configuration management, deployment, and orchestration

  - Source: https://github.com/ansible/ansible

  - Contributors: Over 5K

  - Repository creation date: March 2012

  - Top languages: Python 88%, PowerShell 6.9%

  - License: GPL 3.0

  - Sponsor: Red Hat

  - Popularity: 56K stars, 23K forks, 2K watching

• The Puppet project:

  - Short description: A general administrative management system designed to configure, update, and install systems

  - Source: https://github.com/puppetlabs/puppet

  - Contributors: Over 1K

  - Repository creation date: September 2010

  - Top languages: Ruby 99%

  - License: Apache 2.0

  - Sponsor/Owner: Puppet by Perforce

  - Popularity: 6.8K stars, 2.3K forks, 475 watching

• The Chef project:

  - Short description: A configuration management tool designed to cover the automation of all IT infrastructure

  - Source: https://github.com/chef/chef

  - Contributors: Over 1K

  - Repository creation date: January 2009

- Top languages: Ruby 98%

- License: Apache 2.0

- Sponsor/Owner: Progress Software Corporation

- Popularity: 7.1K stars, 2.6K forks, 374 watching

- The Stackstorm project:

  - Short description: An event-driven automation tool

  - Source: https://github.com/StackStorm/st2

  - Contributors: Over 300

  - Repository creation date: April 2014

  - Top languages: Python 94%

  - License: Apache 2.0

  - Sponsor/Owner: Linux Foundation

  - Popularity: 5.4K stars, 696 forks, 168 watching

- The eNMS automation project:

  - Short description: A higher-level management system to create workflow-based network automation solutions

  - Source: https://github.com/eNMS-automation/eNMS

  - Contributors: 30

  - Repository creation date: October 2017

  - Top languages: Python 53%, JavaScript 26%, HTML 16%

  - License: GLP 3.0

  - Sponsor: N/A

  - Popularity: 700 stars, 148 forks, 73 watching

- NetBrain products:

  - Short description: NetBrain has developed several products for network automation, including **Problem Diagnosis Automation System (PDAS)**

  - Site: https://www.netbraintech.com/

  - Owner: NetBrain Automation

- SolarWinds Network Automation Manager:

  - Short description: Proprietary product developed by SolarWinds for network automation

  - Site: `https://www.solarwinds.com/network-automation-manager`

  - Owner: SolarWinds

Besides the tools and platforms that you can investigate, you also can participate in working groups related to network automation. Let's have a look at some of them.

Joining the network automation community

One of the strategies to improve your knowledge and get updated with new technologies is to participate in the community. The following is a small list of possible groups that you might be interested in watching or participating in:

- IETF netmgmt working group:

  - Short description: A group that has a focus on working on standards for automated network management, such as RESTCONF, NETCONF, and YANG

  - Site: `https://www.ietf.org/topics/netmgmt/`

- Meetup groups:

  - Short description: One good idea is to join a local meetup group that has regular meetings. Then, you can talk with professionals in the same area and improve your network and knowledge. `https://www.meetup.com/` is a site where people can organize and meet.

  - Example in New York: `https://www.meetup.com/Network-to-Coders/`.

  - Example in Sydney: `https://www.meetup.com/it-automation/`.

  - Example in San Francisco: `https://www.meetup.com/sf-network-automation/`.

- **North American Network Operators' Group (NANOG):**

  - Short description: NANOG has tons of documents and presentations, and also organizes conferences where you can find multiple topics on network automation

  - Site: `https://www.nanog.org/`

- **Global Network Advancement Group (GNA-G):**

  - Short description: GNA-G is a community of network professionals from several areas, including research, operation, and education all over the world. They organize meetings and have some documentation resources.

  - Site: `https://www.gna-g.net/`.

- The Network to Code company community:

  - Short description: Network to Code is a consulting company that maintains GitHub repositories and a `slack.com` group that discusses network automation, which can be joined for free

  - GitHub: `https://github.com/networktocode/awesome-network-automation`

  - Slack group: `networktocode.slack.com`

- IP Fabric company community:

  - Short description: The IP Fabric company also maintains GitHub repositories and has a `slack.com` group open to anyone to join

  - GitHub: `https://github.com/community-fabric`

  - Slack group: `ipfabric-community.slack.com`

Other communities also can be found attached to some private companies, such as IBM, Oracle, VMware, Google, and Amazon. They might even use public tools such as Slack, LinkedIn, or GitHub to communicate, and they are maybe more focused on products that these companies offer instead of generic discussion. They are worth checking out, because they might have something to add.

Another idea to improve your knowledge and skills is to contribute to a platform that already exists as a developer, or to build your own if you dare.

I hope this section can give you ideas on the path forward.

Summary

This chapter was focused on getting your hands dirty in a network lab, checking how some code automation works in Go and Python, and, finally, exploring the possibilities of how to go forward with network automation.

At this point, you should be very confident about how to build your own network lab, how to improve your network automation code, and what to do next to continue improving.

On behalf of all the people who worked hard on this book, we want to thank you for investing the time to read this book. It was a hard accomplishment to gather so much information and pass it on to others in an easy and pleasurable way. I hope you have made the most of it by reading it and discovering new technologies and techniques in the realm of network automation.

Now, you can embark on further challenges that will take you deeper into network automation, which you will find when building a complete solution and putting into practice everything that you have learned.

Index

Packtpub.com

Subscribe to our online digital library for full access to over 7,000 books and videos, as well as industry leading tools to help you plan your personal development and advance your career. For more information, please visit our website.

Why subscribe?

- Spend less time learning and more time coding with practical eBooks and Videos from over 4,000 industry professionals

- Improve your learning with Skill Plans built especially for you

- Get a free eBook or video every month

- Fully searchable for easy access to vital information

- Copy and paste, print, and bookmark content

Did you know that Packt offers eBook versions of every book published, with PDF and ePub files available? You can upgrade to the eBook version at packtpub.com and as a print book customer, you are entitled to a discount on the eBook copy. Get in touch with us at customercare@packtpub.com for more details.

At www.packtpub.com, you can also read a collection of free technical articles, sign up for a range of free newsletters, and receive exclusive discounts and offers on Packt books and eBooks.

Other Books You May Enjoy

If you enjoyed this book, you may be interested in these other books by Packt:

Network Automation with Go

Nicolas Leiva, Michael Kashin

ISBN: 9781800560925

- Understand Go programming language basics via network-related examples
- Find out what features make Go a powerful alternative for network automation
- Explore network automation goals, benefits, and common use cases
- Discover how to interact with network devices using a variety of technologies
- Integrate Go programs into an automation framework
- Take advantage of the OpenConfig ecosystem with Go Build distributed and scalable systems for network observability

Mastering Python Networking - Fourth Edition

Eric Chou

ISBN: 9781803234618

- Use Python to interact with network devices
- Understand Docker as a tool that you can use for the development and deployment
- Use Python and various other tools to obtain information from the network
- Learn how to use ELK for network data analysis
- Utilize Flask and construct high-level API to interact with in-house applications
- Discover the new AsyncIO feature and its concepts in Python 3
- Explore test-driven development concepts and use PyTest to drive code test coverage
- Understand how GitLab can be used with DevOps practices in networking

Packt is searching for authors like you

If you're interested in becoming an author for Packt, please visit authors.packtpub.com and apply today. We have worked with thousands of developers and tech professionals, just like you, to help them share their insight with the global tech community. You can make a general application, apply for a specific hot topic that we are recruiting an author for, or submit your own idea.

Share your thoughts

Now you've finished *Network Programming and Automation Essentials*, we'd love to hear your thoughts! Scan the QR code below to go straight to the Amazon review page for this book and share your feedback or leave a review on the site that you purchased it from.

https://packt.link/r/1803233664

Your review is important to us and the tech community and will help us make sure we're delivering excellent quality content.

Download a free PDF copy of this book

Thanks for purchasing this book!

Do you like to read on the go but are unable to carry your print books everywhere? Is your eBook purchase not compatible with the device of your choice?

Don't worry, now with every Packt book you get a DRM-free PDF version of that book at no cost.

Read anywhere, any place, on any device. Search, copy, and paste code from your favorite technical books directly into your application.

The perks don't stop there, you can get exclusive access to discounts, newsletters, and great free content in your inbox daily

Follow these simple steps to get the benefits:

1. Scan the QR code or visit the link below

https://packt.link/free-ebook/9781803233666

2. Submit your proof of purchase

3. That's it! We'll send your free PDF and other benefits to your email directly